Mathematical Modelling of Syst

Mathematical Modelling of Systems and Analysis

K. KAMALANAND
Assistant Professor
Department of Instrumentation Engineering
Madras Institute of Technology Campus, Anna University, Chennai

P. MANNAR JAWAHAR
Vice Chancellor
Karunya Institute of Technology and Sciences
&
Former Vice Chancellor and Senior Professor
Anna University, Chennai

PHI Learning Private Limited
Delhi-110092
2019

₹ 350.00

MATHEMATICAL MODELLING OF SYSTEMS AND ANALYSIS
K. Kamalanand and P. Mannar Jawahar

© 2019 by PHI Learning Private Limited, Delhi. All rights reserved. No part of this book may be reproduced in any form, by mimeograph or any other means, without permission in writing from the publisher.

ISBN-978-81-935938-1-3 (Print Book)
ISBN-978-93-935938-5-1 (e-Book)

The export rights of this book are vested solely with the publisher.

Published by Asoke K. Ghosh, PHI Learning Private Limited, Rimjhim House, 111, Patparganj Industrial Estate, Delhi-110092 and Printed by Rajkamal Electric Press, Plot No. 2, Phase IV, HSIDC, Kundli-131028, Sonepat, Haryana.

Contents

Foreword 1 .... *xi*
Foreword 2 .... *xiii*
Preface .... *xv*

Chapter 1: INTRODUCTION TO THEORY OF SYSTEMS 1–28

1.1 Systems 1
    1.1.1 System Classification 2
1.2 Process 2
1.3 Modelling 3
    1.3.1 Physical Modelling 3
    1.3.2 Mathematical Modelling 3
1.4 Work, Energy and Power 5
    1.4.1 Exergy 10
1.5 Lagrangian Formalism 12
1.6 Hamiltonian Formalism 13
1.7 Energy of a Signal 14
1.8 Disorder and Entropy 16
1.9 Negentropy 19
1.10 Synergy 19
1.11 Euler Angles 20
1.12 Dimensional and Non-dimensional Analysis 20
    1.12.1 Non-dimensionalisation 21
1.13 Mathematical Analogy 23
1.14 Effect of Observation on the System 25
Exercises 27

**Chapter 2: CLASSIFICATION AND REPRESENTATIONS
OF SYSTEMS 29–47**

2.1 Static and Dynamic Systems 30
2.2 Lumped and Distributed Systems 31

v

vi Contents

2.3 Causal and Non-causal Systems..32
2.4 Forced and Unforced Systems...33
2.5 Conservative and Non-conservative Systems 33
2.6 Time-Invariant and Time-Variant Systems......................................34
2.7 Autonomous and Non-autonomous Systems 35
2.8 Hybrid Systems 36
2.9 Zeno Systems 36
2.10 Linear Systems 36
2.11 Non-linear Systems 37
    2.11.1 Parasitic Non-linearity and Voluntary Non-linearity................38
    2.11.2 Typical Physical Non-linearities...................................38
    2.11.3 General Non-linearity..40
2.12 Input and Output Multiplicity 40
2.13 Linearisation 41
    2.13.1 Taylor's Series Method for Linearisation.............................42
    2.13.2 The Jacobian Approach for Linearisation...............................43
2.14 The Describing Function 45
    2.14.1 Steps to Obtain the Describing Function of a Non-linear
            Element 46
Exercises 47

Chapter 3: MATHEMATICAL MODELLING OF SYSTEMS......48–73

3.1 The Regression Line and Correlation Value 49
3.2 Multiple Regression 52
3.3 Polynomial Regression...53
3.4 Modelling Applications in Marketing 54
3.5 Spline Interpolation 55
3.6 Modelling of Dose Concentration 58
3.7 Working with Complex Numbers 59
3.8 Mathematical Modelling using Dimensional Analysis..........................62
3.9 Buckingham's π theorem..64
3.10 Assumptions and Approximations..65
3.11 Modelling Approaches 66
3.12 Continuous and Discrete Models 70
3.13 Applications of Mathematical Models 70
3.14 Dynamic Analysis 72
Exercises 73

Chapter 4: REPRESENTATION OF A FAMILY OF CURVES
            USING ORDINARY DIFFERENTIAL EQUATIONS....74–79

4.1 Family of All Circles..74
4.2 Quadrifolium Curve 77
4.3 Trifolium Curve..78
Exercises 79

Chapter 5: SYSTEM MODELLING USING DIFFERENCE EQUATIONS AND ORDINARY DIFFERENTIAL EQUATIONS 80–97

5.1 Types of Ordinary Differential Equations 80
5.2 Mathematical Modelling of Exponential Growth and Exponential Decay 83
5.3 Mathematical Modelling of Oscillatory Dynamics 84
5.4 Degrees of Freedom 86
5.5 Order of the Differential Equation 87
5.6 Modelling the Motion of Planets and Satellites 88
5.7 Mathematical Modelling using Difference Equations 91
5.8 Integro-differential Equations 94
5.9 Delay Differential Equations 94
5.10 Differential-difference Equations 94
5.11 Fractional Order Differential Equations 95
5.12 Stochastic Differential Equation 96
Exercises 97

Chapter 6: MATHEMATICAL MODELLING USING BALANCE LAWS 98–107

6.1 Mathematical Modelling using Material Balance 98
6.2 Modelling using Energy Balance 99
    6.2.1 Modelling of the Process of Drilling Holes using Laser 99
    6.2.2 Hamilton's Method 100
    6.2.3 The Tautochrone 104
Exercise 107

Chapter 7: MATHEMATICAL MODELLING OF CHEMICAL REACTIONS 108–115

7.1 First Order Reaction 109
7.2 Second Order Reaction 110
7.3 Third Order Reaction 112
7.4 Opposing Reaction 113
7.5 Consecutive Reaction 114
Exercises 115

Chapter 8: MATHEMATICAL MODELLING OF POPULATION DYNAMICS 116–128

8.1 Steady Population 116
8.2 Exponentially Growing or Decreasing Population 116
8.3 Malthusian Model 117

viii Contents

8.4 The Logistic Model..117
8.5 Lotka–Volterra Model..117
8.6 Competition Models..119
8.7 Epidemic Model: The SIR Model for the Spread of Disease.............119
8.8 Modelling of HIV-immune System Interaction.................................121
8.9 Delay Differential Equations in Modelling of Population Dynamics....123
8.10 Population Modelling using Integro-differential Equations................123
8.11 Population Modelling using Difference Equations...........................124
Exercises 127

Chapter 9: MATHEMATICAL MODELLING USING THE CALCULUS OF VARIATIONS 129–141

9.1 Euler's Equation...132
9.2 Variational Problem for a Functional Depending on n Functions......137
9.3 Variational Problem for Functional Dependent on Higher Order Derivatives...138
9.4 Applications of Calculus of Variations...139
Exercises 141

Chapter 10: STABILITY THEORY...142–162

10.1 Fixed Points of a System..142
10.2 Phase Plane Analysis..143
   10.2.1 Classification of Singularities...145
10.3 Liapunov's Direct Method...154
10.4 Potential Functions and Catastrophes...157
   10.4.1 Catastrophe Theory..159
Exercises 161

Chapter 11: STATE SPACE MODELS AND TRANSFER FUNCTIONS...163–177

11.1 The State-Space Model..163
11.2 Controllability and Reachability...167
11.3 Observability..168
11.4 The State Transition Matrix..169
   11.4.1 Properties of State Transition Matrix.................................170
11.5 Various State Space Representations...171
11.6 Transfer Function Models...172
   11.6.1 Transfer Function Models of Fractional Order Systems.........173
   11.6.2 Advantages of State Space Model over Transfer Function Models..173
11.7 Stability Analysis using State Space Models and Transfer Functions......173
Exercises 176

Chapter 12: SYSTEM IDENTIFICATION 178–201

12.1 Experimental Design 178
12.2 Filtering of Measured Input-Output Data 182
12.3 Time Series 182
12.4 Selection of Models 183
12.5 Verification and Validation 194
12.6 The Normal Distribution 194
12.7 Mathematical Modelling using Maximum Entropy Principle (MEP) 194
12.8 Entropy Measures 194
12.9 Procedure for System Identification using Maximum Entropy
 Principle (MEP) 199
Exercises 201

Chapter 13: PARAMETER ESTIMATION TECHNIQUES 202–234

13.1 Estimation of Parameters of Linear Models: Least Squares Method 202
13.2 Parameter Estimation Methods for Non-linear Models 208
  13.2.1 The Extended Kalman Filter 208
  13.2.2 The Unscented Kalman Filter 210
  13.2.3 Dynamic Programming 214
  13.2.4 Maximum Likelihood (ML) Estimation 216
  13.2.5 Estimation of Model Parameters using Evolutionary
   Optimisation and Swarm Intelligence Techniques 218
  13.2.6 Parameter Estimation in Frequency Domain 232
Exercises 234

References *235–237*

Further Reading *239–246*

Index *247–252*

Foreword 1

I am immensely pleased to write the foreword to this book *Mathematical Modelling of Systems and Analysis*. The authors, K. Kamalanand and P. Mannar Jawahar, have taken the readers through the interesting subject of mathematical modelling from fundamentals to advanced topics. The enthusiasm of the authors for the subject is evident in every page of the book.

The book applies an interdisciplinary approach that brings together the knowledge from Mathematics, Mechanics, Biology, Chemistry and other varied disciplines, with the goal to analyse and understand complex problems. Mathematical modelling is one of the essential and challenging tasks faced in various areas. To address this need, the book covers both the fundamental approaches to modelling such as regression modelling and modelling using ordinary differential equations, and also, the contemporary modelling approaches such as neural networks, support vector machines and evolutionary algorithms.

This book is written in such a way that students, academicians and professionals from a variety of domains in science, engineering and technology can easily understand it. Also, since mathematics is the essence of all scientific discoveries and inventions, it will be highly useful for scientists and researchers. I am confident that this book will stimulate the universities engaged in scientific and technological studies and research to remodel their curricula at the undergraduate as well as postgraduate levels provided for the study of modelling techniques. This will enable the students to work in interdisciplinary teams and induce interdisciplinary culture among the departments.

A significant feature of the book is to guide researchers to develop mathematical models for solving problems in science and technology. The approach of the authors to provide several examples and illustrations makes the complex mathematical principles easy to understand. I appreciate the decision of the publisher to bring out this book since mathematical models are utilised in all areas of science and engineering.

Prof. M. Anandakrishnan
PhD, FIE
Padma Shri
Former Vice Chancellor, Anna University
Former Chairman, IIT Kanpur

Foreword 2

It gives me great pleasure to write the foreword to the book *Mathematical Modelling of Systems and Analysis* authored by K. Kamalanand and P. Mannar Jawahar. The book practically gives the readers an interdisciplinary exposure taking them through a plethora of topics like mathematics, mechanics, biology, chemistry and other varied disciplines. The lucid manner of presentation will surely enable the readers to acquire knowledge and to understand and analyse even complex problems effortlessly.

Mathematical modelling is the art of representing the complex dynamics of real systems in simplified mathematical forms. In reality, certain experiments cannot be practically performed. Such experiments can be conveniently simulated using mathematical models. Mathematical modelling is an efficient way of proposing and testing hypothesis. This book takes us on a journey through various mathematical modelling and analysis techniques with examples from a variety of fields in science and technology, including medicine and biology, chemical reactions, mechanical and electrical engineering, population dynamics, etc. The authors have presented the fundamentals of modelling which pave way to the readers to develop their own mathematical models of the system they are working on.

As a physician, I believe that mathematical modelling is a highly significant field, which empowers doctors and technologists to work together on solving a problem of high sociological importance. In the medical domain, mathematical modelling finds its major applications in three important arenas, viz. diagnostics, therapy planning and surgery planning. Predictive mathematical models can be employed to diagnose and screen different diseases using non-invasive physiological measurements. Such models can be greatly handy for mass screening of diseases. Further, mathematical modelling of the dynamics of certain diseases and the interaction of pathogens with the immune system aids us to understand the mechanism and progression of diseases. These models can be utilised for calculating the optimal drug dosages, computing the dose required for destroying cancer cells in patients undergoing radiotherapy and chemotherapy, etc. Also, such mathematical models can be utilized for predicting the medical outcomes for a particular treatment plan. In surgical procedures, mathematical models have a potential to offer useful help to surgeons. For example, in recent years, mathematical models have been developed for predicting the optimal incisions required during surgery and for predicting the impact of surgery on metastasis. Models can also be employed to predict the surgical outcomes and for maximising the efficiency of surgical procedures.

Also, mathematical modelling finds a major role in epidemiology research. The major goal of epidemiological modelling is to understand the interrelationship between the variables that affect the course of infection in an individual and to understand the pattern of infections within communities in a specific geographical location. Such models help us to predict the impact of the epidemic.

This book has simplified the complex mathematical procedures into easily fathomable and thought-provoking material. I believe that this book will be highly beneficial as a guide for researchers and academicians to improve convenient models in a variety of fields and to promote interdisciplinary pathways in research and development.

Dr. Mayilvahanan Natarajan
M.S. Orth., M.Ch. Orth. (L'pool), Ph.D. (Orth., Onco.) D.Sc., F.R.C.S. (Eng)
Padma Shri
Orthopaedic Surgeon
7th Vice Chancellor, The Tamil Nadu Dr. M.G.R. Medical University
1st Director, Institute of Orthopaedics & Traumatology,
Government General Hospital & Madras Medical College
Emeritus Professor, The Tamil Nadu Dr. M.G.R. Medical University

Preface

Mathematical modelling is the art of developing mathematical formulations and equations, which effectively describes or mimics the real-world phenomenon. A mathematical model of a system has to be developed in such a way that it preserves the fundamental properties of the actual system under investigation. Such mathematical models are highly useful for simulation and analysis of the actual system under investigation. Further, certain mathematical systems exhibit the properties of dynamism and adaptation. Electrical engineers utilise these concepts for signal processing, controller design, circuit design and analysis. Computer scientists use these properties for the development of algorithms, design of computers, automata theory, robotics and artificial intelligence. Philosophers use them for analysing causality and strange loops. Biologists use them in the fields of neurosciences, biophysics, ecology and evolution. Artists utilise mathematical models (such as fractal geometry) for generating beautiful art and self-similar patterns. Almost every field of arts, science and engineering utilises mathematical models for the purpose of analysis and application development. It is in the context of mathematical modelling that scientists, engineers, mathematicians and people from various fields can work together for developing systems with high social relevance. The development of mathematical models enables us to analyse the stability of complex physical structures. For example, mathematical modelling can be an important tool in the development of stable drugs, which can decrease bacterial and viral loads.

The contents of this book are structured into thirteen chapters. The first chapter introduces the principles, parameters and features of systems from a classical as well as a modern point of view. Chapter 2 explains various physical and mathematical classifications of systems. Chapter 3 focuses on different mathematical modelling techniques and general procedure for the development of mathematical model of systems. Chapters 4 and 5 provide a general and concrete explanation on mathematical modelling using ordinary differential equations. Chapter 6 explains the development of models based on balance laws such as material balance and energy balance. The mathematical modelling of chemical reactions and population dynamics is explained in Chapters 7 and 8, respectively. Chapter 9 deals with the development of ordinary differential equation (ODE) models using the calculus of variations. Chapter 10 introduces the reader to stability analysis using ODE models. Chapter 11 explains the development of linear models such as transfer functions and state space models. The development of mathematical models using the black box and grey box approaches is explained in Chapter 12. The last chapter, i.e., Chapter 13 focuses on different parameter estimation algorithms for estimating the unknown parameters of linear and non-linear models. The methods of developing mathematical models along with the analysis techniques are extensively presented using examples from several fields such as physics, chemistry, biology, mechanical engineering, electrical engineering, sociology, psychology, etc. Also, several numerical and worked-out examples are presented for the readers' ease of use.

We would like to thank Mrs. R. Padmaja for her help in typing the manuscript, and Mr. K. Srikanth for his help in preparing the images in vector format. It was a pleasure writing this book and watching its development step by step. We hope that this book is useful for the students and researchers from various fields, R&D team of industries and will be enjoyed by the readers.

K. Kamalanand
P. Mannar Jawahar

CHAPTER 1

Introduction to Theory of Systems

Most of the concepts of system theory appear to have evolved from the principles of thermodynamics and classical mechanics. It is important for any engineer or scientist, who wishes to understand and analyse complex physical systems, to have a basic knowledge of thermodynamics, classical mechanics and an understanding of the associated mathematics.

1.1 SYSTEMS

From a general perspective, a *system* can be defined as a group of complex bodies under observation or analysis. In detail, a system can be defined as an organised structure, which consists of a group of bodies that work together in order to achieve the goal of the system. The system is usually limited by physical or natural boundaries. At times, it is quite necessary and convenient to confine a system by means of arbitrary imaginary boundaries. The boundary of a system can be spatial or temporal or spatio-temporal in nature. The remaining part of space forms the surroundings or environment (Figure 1.1). Sometimes, the system may be influenced by its environment.

Before understanding the concepts of system modelling and analysis, one must know the concept of space and time in the study of systems. The concept of space is essential to fix the position of a particle or a system of particles. The position of the particle may be defined by three lengths in the three-dimensional problems or two lengths in the two-dimensional problems. The lengths which are measured from a certain reference point or origin in three or two given directions are known as the *coordinates* of the particle. Further, the concept of time is essential to relate the sequence of events, for example, the starting and stopping of the motion of a system of particles.

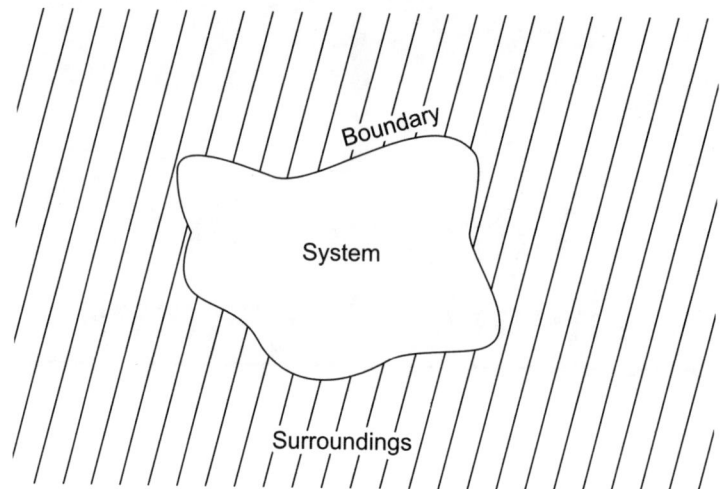

FIGURE 1.1 The system, its boundary and surroundings.

1.1.1 SYSTEM CLASSIFICATION

A system that does not exchange energy or material or information with other systems or its surroundings is called an *isolated or closed system*. Whereas, a system which exchanges energy or material or information with other systems is called an *open system*. Further, if there is no flow of material or energy or information in the system, and its non-variability is not due to any external forces, then the system is said to be in *equilibrium state*. A system is called a *heterogeneous system* if it consists of elements with differing properties or functions and is separated by interfaces. A system having no interfaces is known as a *homogeneous system*.

1.2 PROCESS

Any change in a system connected with a variation in at least one independent parameter or variable such as time and distance is called a *process*.

Further, processes can be classified into reversible and irreversible process. A *reversible process* is a process which when conducted in the reverse direction repeats all the stages of the direct process. An *irreversible process* when conducted in the reverse direction does not repeat the direct process. Consider, for example, a lead acid accumulator or battery, which can be charged by connecting it to another cell with its emf directed oppositely. This process is an example of a reversible process, since the reaction in the cell can proceed in the forward (discharging) or the reverse (charging) direction (Figure 1.2).

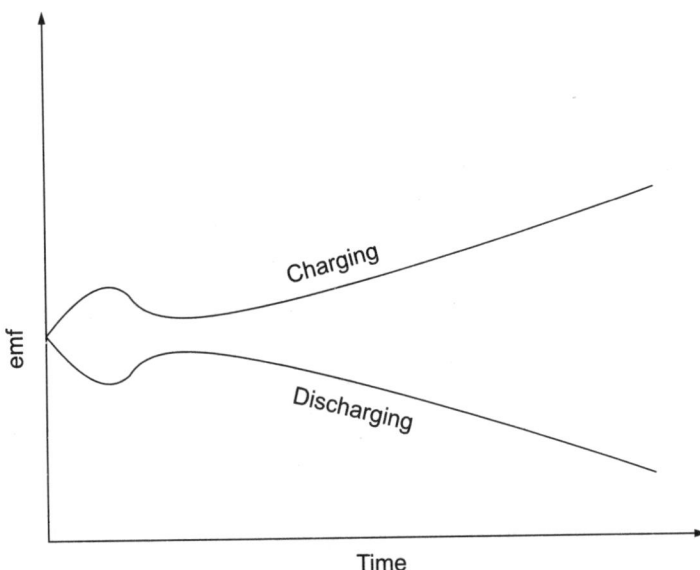

FIGURE 1.2 Charging and discharging of a lead acid battery.

1.3 MODELLING

Modelling is the art of developing a realistic and simplified representation of a system. The developed model of the system is used for the purpose of prediction and analysis. Modelling can be broadly classified as physical modelling and mathematical modelling.

1.3.1 Physical Modelling

Constructing a model physically for a real or actual system is called *physical modelling*. The model may be smaller or tiny in size corresponding to the actual model or prototype model (for example, for a prototype model of a physical automobile vehicle, railway vehicle or aircraft vehicle) or bigger in size corresponding to actual or prototype model (for example, a prototype model of a wrist watch).

1.3.2 Mathematical Modelling

Mathematical modelling is a powerful tool for design and analysis of systems. Mathematical modelling is nothing but converting any physical system into a mathematical form (constructing mathematical equations). These equations may be either algebraic or differential depending on whether the system is in static conditions or in dynamic conditions. In other words, a mathematical model is a simplified mathematical construction of a real system, which enables simulation, and hence, provides a better understanding of the system under analysis. The model is made up of variables (dependent variables, independent variables and parameters) and constants. The structure of the model that maps the inputs to the outputs is known as the *function*.

The development of a mathematical model consists of the following four major steps:

1. Problem formulation: The first step involved in the development of the mathematical model is the identification of the problem. The 'problem' refers to the real-world scenario that needs to be addressed by the model. The structure of the model is based on the problem which is considered.

2. Model formulation: In this stage, the various factors such as the reversibility, the environment or surroundings, system boundaries and transfer of energy are studied. Based on these conditions, the independent variables, dependent variables and the model parameters are selected along with the structure of the model.

3. Study of usefulness of the model: After development of the model, the next step is to test the usefulness and the functionality of the model. In this stage, it is mandatory to check whether the developed model addresses the real-world problem considered.

4. Model validation: The final step involved in mathematical modelling is the validation of the developed mode. In this stage, the outputs of the developed model are compared with the outputs of the actual system and the errors are computed. Based on the error values, the model parameters are fine-tuned.

The second step or the model formulation step is the most complex stage in the modelling process. The contents of this book deal with several methods and procedures involved in the development of the mathematical model of various physical systems and processes.

Further, as the complexity of the system increases, it becomes more difficult to analyse the system. The task of model development involves breaking the complex system into several simple mathematical descriptions so that the engineers and scientists can get more insight into the dynamics of the actual system. Hence, a mathematical model facilitates analysis of complex real-world systems and helps us in understanding the functional and design aspects of that system. Finally, a best model is the simplest model that provides maximum information on the system under analysis. The major objectives of mathematical modelling are as follows:

1. Simulation: Simulation refers to the application of the model to understand the complex dynamics of the system. Due to the recent advances in computers, the simulation of complex mathematical models has become an easy task.

Introduction to Theory of Systems

2. Stability analysis: Stability analysis is an important task in almost all fields of engineering and sciences. Analysis of stability of systems enables us to use the systems more effectively. Further, stability analysis helps us in identifying the range of inputs for maximising the lifetime of the system.

3. Optimisation: Another important application of the mathematical model is to optimise the yield of the actual system. The term 'optimisation' encapsulates several tasks such as maximising profits, making design modifications for maximising performance, design of control systems for operating the systems in desired conditions, etc.

1.4 WORK, ENERGY AND POWER

The work done by a system is due to the interaction between the system and the surroundings as a result of which the external forces violating equilibrium in the system are overcome. The *work* of a process is the energy transmitted by one body to another body upon their interaction. The *energy* of a system is its capacity for doing work. Further, the *power* of a system is defined as the amount of work performed per unit time. The power generated by any system is the amount of work done dW within the time interval dt.

$$\text{Power, } P = \frac{dW}{dt}$$

The energy of a system is generally a function of its state and consists of the following three parts:
1. The kinetic energy of motion of the system
2. The potential energy due to the position of the system
3. The internal energy.

The kinetic energy of a system is the energy which it possesses by virtue of its motion and is measured or expressed by the amount of work that the system can perform against the impressed forces before its velocity is completely destroyed. The kinetic energy is a characteristic of both the translational and rotational motions of the system. In order to analyse problems involving force, velocity and acceleration, it is necessary to develop means of obtaining the body's kinetic energy when the body is subjected to translation, rotation about a fixed axis or general planar motion. Further, the *kinetic energy* of a system is defined as a scalar quantity T equal to the arithmetic sum of the kinetic energies of all the particles of the system, i.e.,

$$T = \Sigma \frac{M_k v_k^2}{2}$$

where, M_k is the point mass of the system and v_k is the velocity of each point of the system.

In translational motion, all the points of the system have the same velocity, which is equal to the velocity of the centre of mass v_c. Therefore, for any point k, we have $v_k = v_c$. Hence, the kinetic energy is represented as follows:

$$T_{trans} = \sum \frac{M_k v_c^2}{2} = \frac{1}{2}(\sum M_k) v_c^2$$

or

$$T_{trans} = \frac{1}{2} M v_c^2 \quad \text{(since } \sum M_k = M\text{)}$$

where, M is the total mass.

In translational motion, the kinetic energy of a body is equal to half the product of the body's mass and the square of the velocity of the centre of mass.

The velocity of any point of a body rotating about an axis is $v_k = \omega h_k$, where h_k is the distance of the point from the axis of rotation and ω is the angular velocity of the body. Hence, the kinetic energy of a rotating system is expressed as follows:

$$T_{rotation} = \sum \frac{M_k \omega^2 h_k^2}{2} = \frac{1}{2}(\sum M_k h_k^2) \omega^2$$

or

$$T_{rotation} = \frac{1}{2} J \omega^2$$

where $J\ (= \sum M_k h_k)$ is the moment of inertia of the body. Hence, in rotational motion, the kinetic energy of a body is equal to half the product of the body's moment of inertia with respect to the axis of rotation and the square of its angular velocity.

Now, consider the motion of a body as a combination of a translation with velocity v_c and a rotation about the axis CP with an angular velocity ω, as shown in Figure 1.3.

The velocity v_k at any point of the body is equal to the geometrical sum of the velocity v_c of the body and the velocity v_k' of the point in its rotation about the axis CP.

$$v_k = v_c + v_k'$$

where, $v_k' = \omega h_k$ and h_k is the distance of the point from the axis CP.

Now,

$$v_k^2 = (v_c + v_k')^2 = v_c^2 + v_k'^2 + 2v_c v_k'$$

Hence, the total energy of the system is given by

$$T_{total} = \frac{1}{2} \sum M_k (v_c^2 + v_k'^2 + 2v_c v_k')$$

$$= \frac{1}{2}(\sum M_k) v_c^2 + \frac{1}{2}(\sum M_k v_k'^2) + \sum M_k v_c v_k'$$

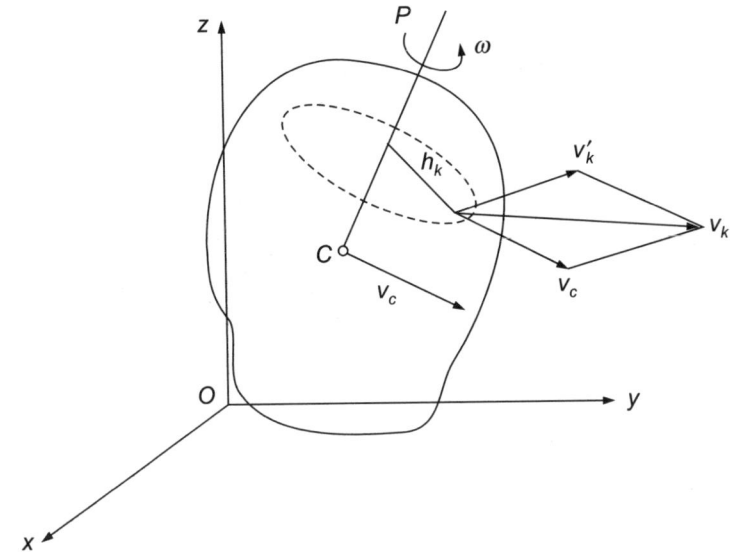

FIGURE 1.3 Body undergoing both translational and rotational motion.

or

$$T_{\text{total}} = \frac{1}{2}(\sum M_k)v_c^2 + \frac{1}{2}(\sum M_k h_k^2)\omega^2 + v_c \sum M_k v_k^2$$

If the work done by a force to move a particle from one point in space to another point is independent of the path followed by the particle, then the force is known as a *conservative force*. For example, the work done by the weight of a particle is always independent of the path, since it depends only on the vertical displacement. Another example of a conservative force is the spring force, where the work done by a spring force is independent of the path and depends only on extension and compression of the spring.

The potential energy of a body is the work it can do by means of its position in passing from its present configuration to zero position (*O*). The *zero position* or the *zero point* is the point at which the capacity to do work is zero. The *potential energy* of a body can also be defined as the amount of work a conservative force will perform when it moves the body from a given position to the reference or zero position. The potential energy can be classified into the following two categories:

1. **Gravitational potential energy:** If a particle is located at a distance y above an arbitrarily selected datum, then the weight of the particle W has positive gravitational potential energy, i.e., $V_g = +Wy$. Similarly, if a particle of weight W has negative gravitational potential energy, $V_g = -W_y$, and it is placed at a distance y below the datum, then $V_g = -W_y$. At the datum, the particle will have zero potential energy $V_g = 0$ (Figure 1.4).

2. Elastic potential energy: When an elastic spring is elongated or compressed to a distance x from its unstretched position, the elastic potential energy V_e due to the spring configuration can be expressed as $V_e = \dfrac{Kx^2}{2}$, where K is the spring constant. Also, a bent spring has potential energy by the work it can do in recovering its natural shape.

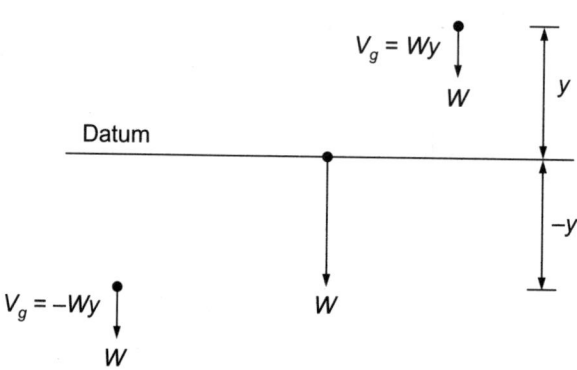

FIGURE 1.4 Positive and negative gravitational potential energy.

In general, the potential energy of a particle in any configuration M is defined as the scalar quantity V equal to the work done on the particle by the forces of a force field in the passage from configuration M to the zero configuration.

$$V = A_{(MO)}$$

The potential energy is dependent on the coordinates of particle M.

$$V = V(x, y, z)$$

Let U be the force function at point M of the field and at the zero point $U_0 = 0$ by assuming that the zero points of the functions $V = (x, y, z)$ and $U = (x, y, z)$ coincide. Now, we have

$$V(x, y, z) = A_{(MO)} = U_0 - U = 0 - U = -U$$

or
$$V(x, y, z) = -U(x, y, z)$$

Hence, the potential energy at any point of a force field is equal to the magnitude of the force function at that point taken with the opposite sign.

By the law of conservation of energy, the total energy remains constant when the body moves from one position to another. In other words, the sum of a particle's potential energy and kinetic energy is constant through the motion of the particle.

$$T_1 + V_1 = T_2 + V_2$$

where, T and V represent the kinetic and potential energy of the system, respectively. In the case of a pendulum, the potential energy which the bob possesses when at rest in its highest position becomes converted into kinetic energy as the bob swings down to its lowest position and is reconverted into potential energy as the bob travels to its next highest position.

Consider the free fall of a particle from rest at a height h to the zero point or an arbitrary datum point. Let x be the instantaneous height of the falling particle from the zero point, the kinetic energy of the particle is

$$T = \frac{1}{2}mv^2$$

where m is the mass of the particle and v is the velocity of the particle.

By definition, $T = mgx$.

Also, the potential energy of the particle is

$$V = mg(h - x)$$

Total energy $= T + V = mgx + mgh - mgx = mgh$

Hence, the sum of potential and kinetic energy at any point is constant.

In order to understand the concept of internal energy, consider a cyclic thermodynamic process, where Q is the heat and W is the work done by the system. In this case, we have

$$Q = W$$

or

$$Q - W = 0 \tag{1.1}$$

According to Eq. (1.1), it is clear that it is not possible to increase the amount of energy in an isolated system. In other words, this explains that a perpetual motion machine of the first kind is impossible to be built.

Now, consider a non-cyclic process, where we have $Q \neq W$, because apart from the conversion of heat into work, a change occurs in the system itself. For a non-cyclic process the quantity $Q-W$ will not depend on the path of transition, but is determined only by the initial and final states of the system. Hence, the quantity $Q-W$ for any process is determined by the change in a certain property of the system, which is known as the *internal energy*.

For a stationary system, in the absence of an external force field, the total energy of the system will equal the internal energy. The internal energy of a system is the sum of

1. Kinetic energy of molecular motion (transitional and rotational)
2. The intermolecular energy (the energy of mutual attraction and repulsion of the particles forming the system)
3. The intramolecular energy (chemical energy)
4. The energy of electronic excitation
5. Intranuclear energy

6. Radiant energy
7. Gravitational energy due to the attraction of particles of a substance to one another.

The change in internal energy of a system undergoing a non-cyclic process is denoted by $\Delta U = U_2 - U_1 = Q - W$. However, for a cyclic process, $\Delta U = 0$.

1.4.1 EXERGY

Exergy is the part of energy that is useful and available. A system in complete equilibrium with its surroundings does not have any exergy. The exergy of a system increases as the system moves away from equilibrium. For example, a block of ice has more exergy in summer when compared to winter and hot water has higher exergy in winter when compared to summer. The exergy transfer associated with heat transfer depends on the temperature at which it occurs in relation to the temperature of the environment. Hence, exergy depends on the state of the system as well as on the state of the surroundings.

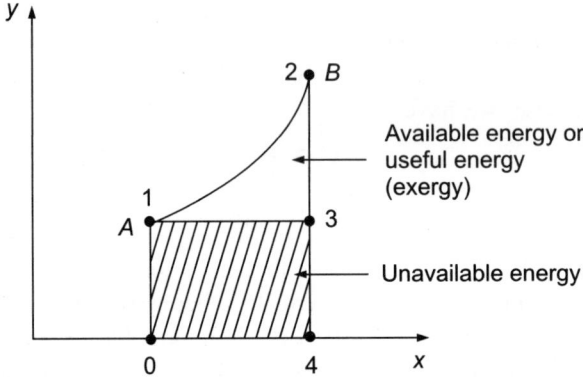

FIGURE 1.5 The exergy of a system.

Consider a process given by $A \rightarrow B$, represented graphically in Figure 1.5. In this process, the available or useful energy is represented by the area under the curve enclosed by 1-2-3-1 and the unavailable energy is represented by the area 0-1-3-4-0. The available energy or the useful energy given by $Area_{1-2-3-1}$ is known as the *exergy* of the system. The concepts of energy, exergy and entropy rise from the concepts of thermodynamics, but are applicable to all fields of science and engineering.

Now, consider the work done by a system. A *work* is said to be done if a system undergoes a change due to the application of an external forcing function. For example, consider the work done by an elastic spring due to an applied force. The magnitude of force developed in a linear elastic spring which is displaced by a distance x from its original position is $F = Kx$, where K is the stiffness of the spring. The work done by the spring is always positive when it is compressed or elongated [Figure 1.6(a)]. The work done W is given by

$$W_{1-2} = \int_{x_1}^{x_2} Kx\, dx$$

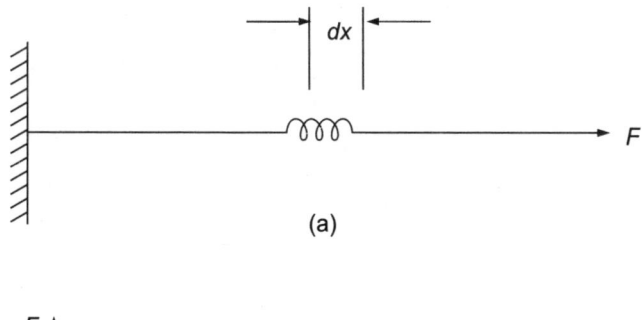

$$W_{1-2} = K\left[\frac{x^2}{2}\right]_{x_1}^{x_2} = \frac{1}{2}[Kx_2^2 - Kx_1^2]$$

The work done can also be calculated by plotting a graph of force versus displacement [Figure 1.6(b)]. The area of the trapezium will give the work done.

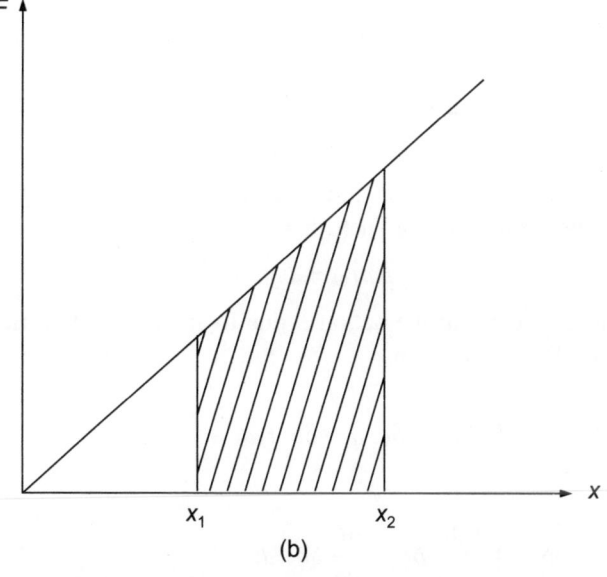

FIGURE 1.6 Work done by elastic spring.

Further, according to the principle of work and energy, the sum of a system's initial kinetic energy and the work done by all the external forces

12 *Mathematical Modelling of Systems and Analysis*

and couple moments acting on the system as the system moves from its initial to final position, is equal to the system's final kinetic energy, i.e.,

$$T_1 = \sum W_{1-2} = T_2$$

1.5 LAGRANGIAN FORMALISM

A system which demands knowledge of not less than N variables for determining the position and orientation of that system is known as a system with N degrees of freedom. Each of the N variables relates to the location and orientation of the system in the physical space. Further, there are only N generalised coordinates associated with the system. Now, consider a system with N generalised coordinates, $(q_1, q_2, ..., q_N) \equiv q$. The potential energy V of the system is a function of q and the kinetic energy of the system is a function of both q and its derivative $\dot{q}$. Then, the Lagrangian of the system is given by

$$L = T(q, \dot{q}) - V(q)$$

If a system moves between two points $1 = [q^{(1)}, t_1]$ and $2 = [q^{(2)}, t_2]$, then the motion of the system follows:

$$\delta \int_1^2 L(q, \dot{q}, t)\, dt = 0$$

Let $q(t)$ be the motion that makes the integral $S = \int_1^2 L\, dt$ an extremum.

Now, if q is given a small perturbation

$$q(t) + \delta q(t)$$

where, $\delta(q)$ is an arbitrary function, which is very small in the interval $t = [t_1, t_2]$, and the variations are such that

$$\delta q(t_1) = \delta q(t_2) = 0$$

Then, it is stated that the action S due to such a small variation vanishes as $q(t)$ makes S an extremum.

$$\delta S = \int_1^2 L(q + \delta q, \dot{q} + \delta \dot{q}, t)\, dt - \int_1^2 L(q, \dot{q}, t)\, dt$$

$$\delta S = \int_1^2 \left(\frac{\partial L}{\partial q} \delta q + \frac{\partial L}{\partial \dot{q}} \delta \dot{q} \right) dt$$

$$\delta S = \left[\frac{\partial L}{\partial \dot{q}} \delta q \right]_1^2 + \int_1^2 \left[\frac{\partial L}{\partial q} - \frac{d}{dt}\left(\frac{\partial L}{\partial \dot{q}} \right) \right] \delta q\, dt$$

Since $\delta(q)$ is small, it is necessary that

$$\frac{d}{dt}\left(\frac{\partial L}{\partial \dot{q}_i}\right) - \frac{\partial L}{\partial q_i} = 0, \quad i = 1, 2, 3, ..., N \tag{1.2}$$

Equation (1.2) is known as the *Euler–Lagrange equation*. For modelling of systems using the calculus of variations (Chapter 9), this equation finds a variety of applications.

For a system containing several particles, the kinetic energy, of each particle in 3 dimensional co-ordinate system (x, y, z) can be given as $T = \frac{1}{2}m(\dot{x}^2 + \dot{y}^2 + \dot{z}^2)$, where $\dot{x}, \dot{y}, \dot{z}$ are the derivatives of x, y, z. Formally, for obtaining the kinetic energy in a new representation (ζ, η, ξ), we require Eqs. (1.3), (1.4) and (1.5) for the transformation.

$$\zeta = \zeta(x, y, z, t) \tag{1.3}$$

$$\eta = \eta(x, y, z, t) \tag{1.4}$$

$$\xi = \xi(x, y, z, t) \tag{1.5}$$

The Lagrangian must be constructed by choosing the set of variables which are closely associated with the dynamics and the geometry of the system.

1.6 HAMILTONIAN FORMALISM

Suppose L is not a function of q_i. Then, $\frac{\partial L}{\partial \dot{q}_i}$ is a constant and does not change with time. Then, the coordinate q_i is cyclic. Now, we change the variables from $(q, \dot{q})$ to $\left(q, \frac{\partial L}{\partial \dot{q}}\right)$, i.e., we transform $L(q, \dot{q}, t)$ into the Hamiltonian $H\left(q, \frac{\partial L}{\partial \dot{q}}, t\right)$, where, $H = \sum_{i=1}^{N} \frac{\partial L}{\partial \dot{q}_i}\dot{q}_i - L$ and $\frac{\partial L}{\partial \dot{q}_i} = p_i$ which is the momentum conjugate to the coordinate q_i and is known as the *canonical momentum*.

Then, $\qquad H = H(q, p, t)$

Now, finding the differential of H, and using the Lagrangian equations, we get

$$dH = \sum_{i=1}^{n} \dot{q}_i dp_i - \sum_{i=1}^{n} \dot{p}_i dq_i - \frac{\partial L}{\partial t} dt$$

which follows that

$$\dot{q}_i = \frac{\partial H}{\partial p_i} \tag{1.6}$$

$$\dot{p}_i = -\frac{\partial H}{\partial q_i} \qquad (1.7)$$

$$\frac{\partial H}{\partial t} = -\frac{\partial L}{\partial t} \qquad (1.8)$$

Equations (1.6), (1.7) and (1.8) are known as the *Hamiltonian equations,* and are used to efficiently find the equations of motion of a system.

For conservative systems whose coordinates are not dependent on time, the Hamiltonian is the total energy $T + V$ and is a constant.

Suppose if the system is conservative and the constraints are not time-dependent, then the Lagrangian L will not have the time explicitly. Hence,

$$\frac{dL}{dt} = \sum_{i}^{N} \left(\frac{\partial L}{\partial q_i} \frac{dq_i}{dt} + \frac{\partial L}{\partial \dot{q}_i} \frac{d\dot{q}_i}{dt} \right)$$

and using the Euler–Lagrange equation, we get

$$\frac{dL}{dt} = \sum_{i}^{N} \left[\frac{d}{dt}\left(\frac{\partial L}{\partial \dot{q}_i}\right) \dot{q}_i + \frac{\partial L}{\partial \dot{q}_i} \frac{d\dot{q}_i}{dt} \right]$$

$$= \sum_{i}^{N} \frac{d}{dt}\left(\dot{q}_i \frac{\partial L}{\partial \dot{q}_i} \right)$$

or

$$\frac{d}{dt}\left(L - \sum_{i}^{N} \dot{q}_i p_i \right) = 0$$

Hence, H is a constant.

If the potential V is not a function of velocity and is only a function of position, then $p_i = \dfrac{\partial L}{\partial \dot{q}_i} = \dfrac{\partial T}{\partial \dot{q}_i}$, and hence,

$$\frac{d}{dt}\left(L - \sum_{i}^{N} \dot{q}_i \left(\frac{\partial T}{\partial \dot{q}_i} \right) \right) = 0$$

1.7 ENERGY OF A SIGNAL

Consider a mass-spring system whose motion is defined by the second order differential equation.

$$\frac{d^2 x}{dt^2} + \frac{K}{M} x = 0$$

where, x is the displacement, K is the spring constant and M is the mass. Its solution is the simple harmonic motion given by

$$x(t) = A\cos(\omega t + \phi) \tag{1.9}$$

where, A is the amplitude of the oscillation, $\omega = \sqrt{K/M}$ is the frequency of the oscillation and ϕ is the initial phase. The total energy of the system is actually the sum of the potential energy and the kinetic energy, which is given by

$$\text{Total energy, } E = \frac{1}{2}Kx^2 + \frac{1}{2}M\dot{x}^2 \tag{1.10}$$

Substituting Eq. (1.1) in Eq. (1.2), we get

$$E = \frac{1}{2}M\omega^2 A^2$$

or

$$E \propto \omega^2 A^2$$

Hence, the energy of an oscillatory sequence is proportional to the square of the amplitude, and also, to the square of the frequency of oscillation.

Based on Teager's and Kaiser's observations (Kaiser 1990; Kaiser 1993), the energy of a signal can be efficiently expressed using the *Teager–Kaiser Energy (TKE) operator* (ψ). It defines the energy required to produce the signal and is useful for viewing and analysing signals from an energy point of view (Li et al., 2007).

Let x_n be the sampled signal generated by a dynamic system whose energy is to be analysed. The discrete TKE operator of x_n is defined as

$$\psi_d[x_n] = x_n^2 - x_{n+1}x_{n-1}$$

For example, consider a signal given by

$$x_n = A\cos[\omega n + \phi]$$

where, n is the sample index, A is the amplitude, ω is the angular frequency, and ϕ is the initial phase.

Now,

$$x_{n+1} = A\cos[\omega(n+1) + \phi]$$
$$x_{n-1} = A\cos[\omega(n-1) + \phi]$$
$$x_{n-1}x_{n+1} = A^2\cos[\omega(n-1) + \phi]\cos[\omega(n+1) + \phi]$$

Using trigonometric identities,

$$\cos(\alpha + \beta)\cos(\alpha - \beta) = \frac{1}{2}[\cos(2\alpha) + \cos(2\beta)]$$

and

$$\cos 2\alpha = 1 - 2\sin^2\alpha = 2\cos^2\alpha - 1$$

We have

$$x_{n-1}x_{n+1} = \frac{A^2}{2}[\cos(2\omega n + 2\phi) + \cos(2\omega n)]$$

$$= \frac{A^2}{2}[2\cos^2(\omega n + \phi) - 1 + 1 - 2\sin^2(\omega n)]$$

$$= A^2\cos^2(\omega n + \phi) - A^2\sin^2(\omega n)$$

$$= x_n^2 - A^2\sin^2(\omega n)$$

Therefore the TKE operator is

$$\psi_d[x_n] \approx A^2\sin^2[\omega n]$$

Hence, it is seen that the TKE operator is proportional to the instantaneous amplitude A and frequency ω of the given signal.

In the continuous domain, the Teager–Kaiser energy operator is defined as

$$\psi_c[x(t)] = \left(\frac{dx(t)}{dt}\right)^2 - x(t)\frac{d^2x(t)}{dt^2}$$

1.8 DISORDER AND ENTROPY

In a spontaneous, irreversible change, the disorder always increases. For example, buildings or even entire civilizations over time are reduced to a pile of dust. However, the dust can never reorganise itself into a building. Consider a case with black and white balls arranged so that the balls of each colour occupy a separate part of the case in an ordered fashion [Figure 1.7(a)]. If we now shake the case a few times, the balls in it move around randomly and become arranged in a state of complete disorder [Figure 1.7(b)]. We cannot, however, restore order in the case by shaking it so that the balls of the same colour would again occupy separate parts of the case. In this context, entropy is a measure of disorder of a system.

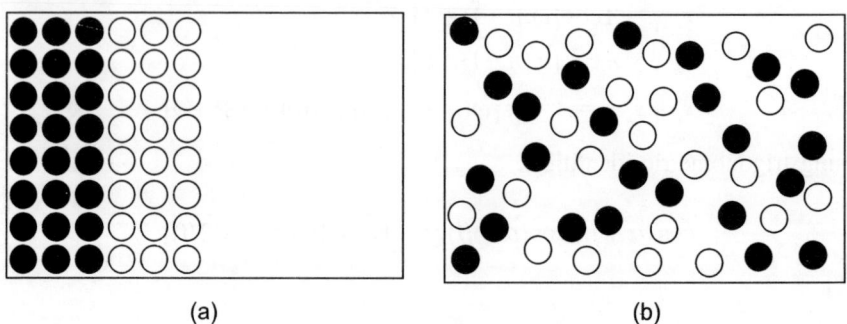

FIGURE 1.7 An illustration for entropy of a system.

The *macrostate* of a system is a state which may exist with a varying distribution of the energy between separate molecules or particles. In a macrostate, there is a large number of microstates, provided that the mean energy is constant. The thermodynamic probability is the number of microstates by which a given macrostate can be created. All microstates are considered to be equally probable, and in a course of time, a system must pass through all the microstates corresponding to a given macrostate. The state of every molecule is determined by three constituent positions (the axis of spatial coordinates x, y and z) and three components of motion (the product of the mass and the velocity vector mv_x, mv_y and mv_z). Thus, the space can be viewed as six-dimensional space called the *gamma space,* which characterises the state of a system at any moment. The gamma space can be divided into a number of cells each having the edges dx, dy, dz, $d(mv_x)$, $d(mv_y)$ and $d(mv_z)$. The volume of a cell equals $dx.dy.dz.d(mv_x).d(mv_y).d(mv_z)$. A given phase cell contains molecules whose coordinates are confined within the limits from x to $x + dx$, from y to $y + dy$, from z to $z + dz$, ... It is now possible to distribute all the molecules of a system according to the values of these coordinates between the corresponding cells of the gamma space, and the distribution of the molecules permits us to find the number of microstates corresponding to a given macrostate, which is the thermodynamic probability.

The thermodynamic probability of a state can be determined by the number of permutations of the number of particles present at a time at which a given spatial distribution occurs. If the given system contains N molecules, then the total number of permutations is $N!$. The thermodynamic probability can be defined as

$$P = \frac{N!}{N_1! N_2! N_3! ...}$$

where, $N = N_1 + N_2 + N_3 + ...$

For example, consider a 15-molecule system whose distribution is shown in Figure 1.8, where each cell is the phase cell.

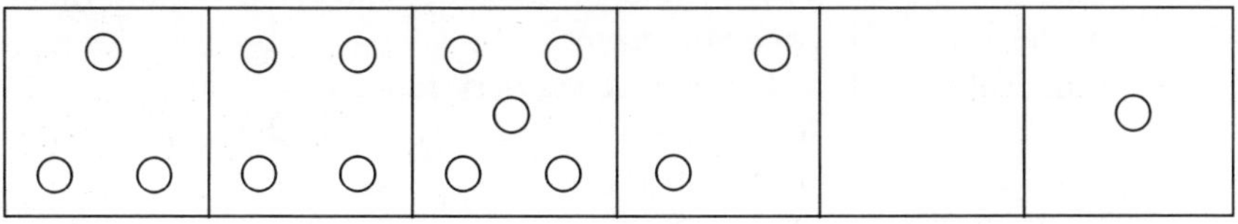

FIGURE 1.8 Typical distribution of molecules in the phase space.

The thermodynamic probability is given as $P = \dfrac{15!}{3!4!5!2!1!} = 37837800$.

Hence, the given macrostate corresponds to 37837800 different microstates.

Hence, the value P is the number of available states or the number of distinguishable microstates that correspond to a given macrostate. In a system with M available states, we measure the entropy of that system using a function S, where $S = K \ln P$ (where, $K = 1.38 \times 10^{-23}$) is the Boltzmann constant.

All assemblies of M will spontaneously evolve in such a way so as to maximise the total entropy of the assembly and its surroundings. The final steadystate reached by an isolated or closed assembly will have the maximum entropy. Hence, if a system is isolated, then its entropy does not change when reversible process occurs in it and its entropy increases when irreversible processes take place. Further, a process cannot happen in an isolated system if it is associated with a decrease in entropy. In other words, the irreversibility of a process can, therefore, be measured by the entropy.

This is actually the second law of thermodynamics. If a system is in a state with a given entropy, it should be expected with an overwhelming probability to pass over to a state with a greater entropy. Hence, all naturally occurring processes proceed so that the entropy of the system grows, tending to a maximum value. The maximum value of entropy is reached when the system arrives at a state of equilibrium.

It is obvious that the degree of disorder is highly similar and correlated to its entropy. Hence, entropy is considered as a measure of the disorder of the system. Due to the relationship between the entropy and the probability of a state, we can say that a state with greater disorder is characterised by greater thermodynamic probability than a more ordered state.

In order to have a better insight and understanding of the concepts of entropy, the following points are presented from the Austrian physicist, Erwin Schrödinger's book, "What is life?", first published in 1944.

1. A system is alive when it does something, moves, exchanges material with its environment.
2. When a system is dead and is isolated or placed in a uniform environment, all motions come to rest as a result of various kinds of friction and temperature becomes uniform by heat conduction.
3. A dead system is an inert lump of matter where a permanent state is reached and in which no observable events occur.
4. Such a state is known as the *state of thermodynamic equilibrium* or the *state of maximum entropy*.
5. Practically, such a state is reached very rapidly.
6. Theoretically, such a state is often not yet an absolute equilibrium nor the true maximum entropy state.
7. Every process, event, happening or anything that is going on in nature means an increase in the entropy of the part of the world where it is going on.
8. A living organism continuously increases its entropy or produces positive entropy, and thus, tends to approach the state of maximum entropy, which is death.
9. An organism, in order to be alive, needs to continuously draw negative entropy from its environment.
10. Entropy = $K \log(D)$ where K is the Boltzmann constant and D is a quantitative measure of the atomic disorder of the system.

EXAMPLE 1.1 Which one has higher entropy—a container containing water or a container containing chipped ice?

Solution: Water would have more entropy or disorder than chipped ice, since the molecules of water have no fixed location. The molecules of ice have fixed locations, and hence, have less entropy or disorder when compared to water. The first case is, however, more stable, whereas the second case is unstable, since ice has the tendency to melt.

> The total energy in the universe is constant; the total entropy is continually increasing.
>
> —*Rudolf Clausius*

1.9 NEGENTROPY

A living organism delays its death by feeding on negative entropy. In other words, a living system tends to keep its entropy low by the process of eating, drinking, breathing, assimilating and metabolising, and hence, maintains itself on a stationary and low entropy level.

If D is a measure of disorder, its reciprocal ($1/D$) can be regarded as a direct measure of order. Thus,

$$-(\text{Entropy}) = K \log\left(\frac{1}{D}\right)$$

This is known as the *negative entropy* or *negentropy* or *syntropy*, and is a measure of order.

1.10 SYNERGY

Synergy refers to the phenomenon where the combination of two or more individual members outperforms the best individual member. In other words, synergy refers to the improved performance of systems that collaborate with each other. This phenomenon is often explained with the phrase "two plus two is equal to five". For example, consider the case where two or more drugs interact with each other to enhance the effects (positive synergy) or side effects (negative synergy) of the individual drugs. This enhanced effect which is greater than the sum of the individual effects of each drug when used separately is known as the *synergetic effect*. Synergy can often be observed in a variety of natural and artificial systems. Consider a plum cake, which is made using butter, eggs, flour, sugar and plums. The taste of the cake is greater than the taste of individual materials used to make the cake. Consider another example of an automobile, which is constructed using a variety of components such as springs, levers, screws, clamps, etc. The assembly of such individual components creates a more useful structure, which performs a variety of useful functions.

An electrical power plant alone has a low efficiency and a conventional water heater when used alone has a low efficiency. In both the cases, the unused energy goes to waste (entropy). However, by combining the processes into one system, energy efficiency as high as 95% can be achieved. This is another classical synergetic effect.

1.11 EULER ANGLES

The altitude and trajectory of a moving body are defined in accordance with the orthogonal axis system fixed on the earth. The coordinates are X-coordinate or axis, Y-coordinate and Z-coordinate representing the forward travel (longitudinal horizontal axis), travel to the right and the vertical travel, respectively. The relation between the body's coordinate system (xyz) and the earth's fixed coordinate system (XYZ) is defined by Euler angles. The Euler angles are represented by a sequence of three angular rotations. *Yaw* denotes the angular oscillation about the z-axis, *pitch* represents the angular oscillation about the y-axis and *roll* represents the angular oscillation about the x-axis (Figure 1.9).

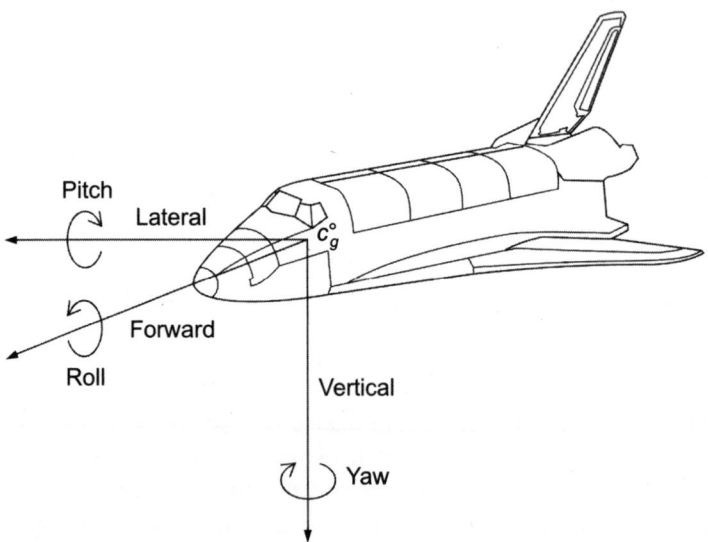

FIGURE 1.9 Yaw, pitch and roll.

1.12 DIMENSIONAL AND NON-DIMENSIONAL ANALYSIS

The principle of dimensional homogeneity states the basic requirements for any mathematical model composed of algebraic and differential equations, which describes a real-world phenomenon to be meaningful, i.e., the left-hand side and the right-hand side of these equations must have the same dimensions. Mass (M), length (L) and time (T) are regarded as fundamental dimensions along with temperature (θ), and all other dimensions are considered as derived dimensions. The dimensions of some common physical quantities are presented in Table 1.1.

Now, consider the Bernoulli equation, which is a dimensionally homogeneous equation. The standard form of the Bernoulli equation for incompressible and irrotational fluid flow is given by

$$P + \frac{1}{2}\rho V^2 + \rho g z = C$$

where, P is the pressure, ρ is the density, V is the velocity, g is the gravitational acceleration and h is the elevation. We can prove that all the additive terms of the equation have the same dimensions (of pressure).

$$\{P\} = \{\text{Pressure}\} = \left\{\frac{\text{Force}}{\text{Area}}\right\} = \left\{\text{Mass}\frac{\text{Length}}{\text{Time}^2}\frac{1}{\text{Length}^2}\right\} = \left\{\frac{M}{T^2 L}\right\}$$

$$\left\{\frac{1}{2}\rho V^2\right\} = \left\{\frac{\text{Mass}}{\text{Volume}}\left(\frac{\text{Length}}{\text{Time}}\right)^2\right\} = \left\{\frac{\text{Mass}}{\text{Length}^3}\frac{\text{Length}^2}{\text{Time}^2}\right\} = \left\{\frac{M}{T^2 L}\right\}$$

$$\{\rho g z\} = \left\{\frac{\text{Mass}}{\text{Volume}}\frac{\text{Length}}{\text{Time}^2}\text{Length}\right\} = \left\{\frac{\text{Mass}}{\text{Length}^3}\frac{\text{Length}^2}{\text{Time}^2}\right\} = \left\{\frac{M}{T^2 L}\right\}$$

By the law of dimensional homogeneity, the constant C in the right-hand side of the Bernoulli equation must have the same dimensions as that of the left-hand side. Hence, the primary dimension of constant C is $\left\{\frac{M}{T^2 L}\right\}$. A mismatch in the dimensions of any of the terms of the equation is an indicator of some error in the equation.

There are a number of important applications of the principle of dimensional homogeneity. For example, it can be used for correcting errors in formulas or equations.

1.12.1 Non-dimensionalisation

In order to efficiently analyse complex processes, in some cases, it is necessary to transform the system of physical quantities such as mass, energy, momentum, etc., and the balance equations into a dimensionless form. The main reason of non-dimensionalisation of equations is that it is possible to obtain a generalisation of the results of theoretical and experimental investigations. Also, numeric calculations can be easily performed using non-dimensional governing equations, since computational complexity is greatly

TABLE 1.1   Dimensions of Some Common Physical Quantities

Quantity	Dimension
Mass	M
Length	L
Time	T
Frequency	T^{-1}
Area	L^2
Volume	L^3
Velocity	LT^{-1}
Acceleration	LT^{-2}
Force	MLT^{-2}
Pressure	$ML^{-1}T^{-2}$
Momentum	MLT^{-1}
Angular momentum	ML^2T^{-1}
Moment of inertia	ML^2
Density	ML^{-3}
Energy/work/heat	ML^2T^{-2}
Power	ML^2T^{-3}
Entropy	ML^2T^{-2}
Kinematic viscosity	L^2T^{-1}
Dynamic viscosity	$ML^{-1}T^{-1}$
Torque	ML^2T^{-2}
Stress	$ML^{-1}T^{-2}$
Strain	—
Young's modulus	$ML^{-1}T^{-2}$
Spring constant	MT^{-2}
Gravitational constant	$M^{-1}L^3T^{-2}$
Gravitational potential	L^2T^{-2}
Temperature	θ
Coefficient of thermal conductivity	$MLT^{-3}\theta^{-1}$
Universal gas constant	$ML^2T^{-2}\theta^{-1}$
Current	$A = M^{1/2}L^{1\frac{1}{2}}T^{-2}$
Charge	TA
Electric potential	$ML^2T^{-3}A^{-1}$
Capacitance	$M^{-1}L^{-2}T^4A^2$
Resistance	$ML^2T^{-3}A^{-2}$
Magnetic induction	$MT^{-2}A^{-1}$
Magnetic flux	$ML^2T^{-2}A^{-1}$

reduced. Further, dependencies between dimensionless parameters can be easily obtained. Another reason for non-dimensionalisation of variables and constants is to overcome the problem of modelling complex processes where the actual conditions of the problem cannot be determined. The procedure for non-dimensionalisation of model equations is performed by transformation of the model equation given by

$$f(X, \theta, t) = 0 \qquad (1.11)$$

where, $[X, \theta, t]$ includes differential operators, independent variables as well as constants. These terms are dimensional in nature. The variables and constants included in Eq. (1.11) can be rendered dimensionless by using some characteristic scales. Then, the dimensionless variables and constants of the problem are expressed as

$$\bar{X} = \frac{X}{X_*}, \quad \bar{\theta} = \frac{\theta}{\theta_*}, \quad \bar{t} = \frac{t}{t_*}$$

where, the terms X_*, θ_*, t_* denote the characteristic scales and are known as *dimensional multipliers*. The dimensionless variables and parameters are denoted by $\bar{X}, \bar{\theta}, \bar{t}$. The dimensionless equations contain a number of dimensionless groups, which represent themselves as some combinations of the physical properties of the system. The physical meaning of these groups is determined by a specific situation as well as by a particular model used for describing the physical phenomena characteristic of that situation. For example, consider the Reynolds number which belongs to a dimensionless group. The Reynolds number is the ratio of the inertia force (F_i) to the friction force (F_f) whose units get cancelled out rendering the Reynolds number dimensionless. Further, if the non-dimensional terms in an equation are of order unity, the equation is called a *normalised equation*.

1.13 MATHEMATICAL ANALOGY

A *mathematical analogy* between two different physical systems can be developed if the mathematical structures of the given systems are identical. Such an analogy preserves most of the features of the structures under analysis. For example, sometimes, mechanical systems are analysed using equivalent electrical circuits since, it is easier to obtain experimental results from electrical circuits than from mechanical systems. The electrical analogy of a mechanical system is obtained by writing and comparing the governing equations of both the systems and making them mathematically similar. The fundamental ideology for such an analogy is that the equations of motion for mechanical systems and electrical systems are identical in structure. Hence, an electrical circuit can easily be generated, which is analogous with a mechanical system whose equations of motion are identical. Further, by analysing the voltage or current in the network, the dynamics of the original mechanical system can be predicted.

24 Mathematical Modelling of Systems and Analysis

There are two possible electrical analogies of mechanical systems, which are as follows:

1. The force-current analogy
2. The force-voltage analogy

To understand the force-current analogy, consider the mechanical and electrical systems shown in Figure 1.10(a) and 1.10(b), respectively. The governing equation of the mechanical system is

$$M\ddot{x} + B\dot{x} + Kx = F(t) \tag{1.12}$$

where, M is the mass, B is the damping coefficient, K is the spring constant, F is the applied force, x is the displacement, $\dot{x} = \dfrac{dx}{dt}$ and $\ddot{x} = \dfrac{d^2x}{dt^2}$.

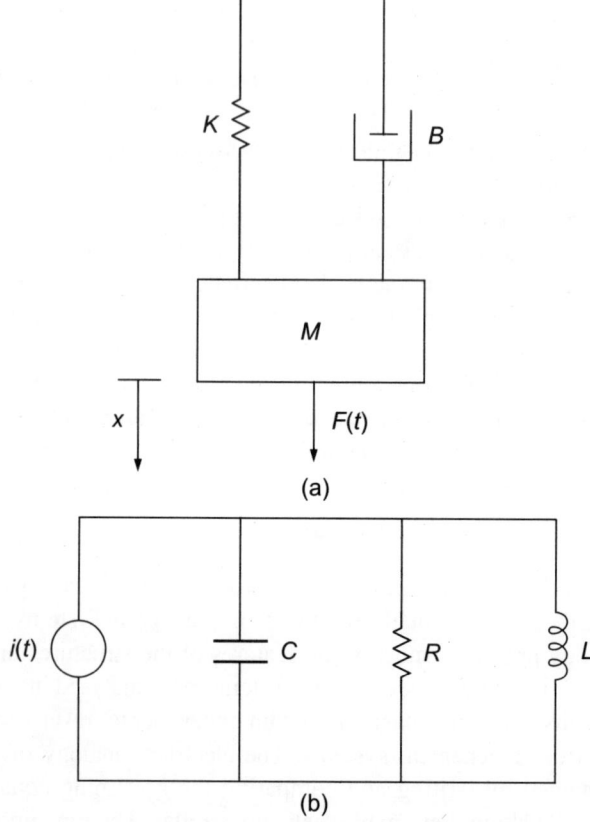

FIGURE 1.10 (a) The mechanical system and (b) Its equivalent electrical network.

The equation for current in the electrical system is

$$C\ddot{v} + \frac{1}{R}\dot{v} + \frac{1}{L}v = \frac{di}{dt} \tag{1.13}$$

where, C is the capacitance, R is the resistance, L is the inductance, i is the current, v is the voltage, $\dot{v} = \dfrac{dv}{dt}$ and $\ddot{v} = \dfrac{d^2v}{dt^2}$.

Integrating Eq. (1.13), we get

$$C\dot{v} + \frac{1}{R}v + \frac{1}{L}\int v\,dt = i(t) \tag{1.14}$$

Comparing Eqs. (1.12) and (1.14), we obtain the force-current analogy, where the integral of voltage is analogous with the displacement x. Hence, in this case, force is analogous with current and velocity is analogous to voltage. In a similar way, it is possible to obtain the force-voltage analogy where force is analogous to voltage and velocity is analogous to current.

1.14 EFFECT OF OBSERVATION ON THE SYSTEM

The development of a mathematical model and analysis of any system requires the act of observation or measurement to be performed on the system. However, the act of observation or measurement itself can affect or influence the system itself. This theory can be explained by the Schrödinger's thought experiment, known as the *Schrödinger's cat* devised by Erwin Schrödinger in 1935.

The experiment can be stated as follows:

A cat, a flask of poison, a radioactive source and a Geiger Muller counter are placed in a sealed box. The arrangement is made in such a way that the observer is not able to look inside the box in the middle of the experiment. If the internal monitor (i.e., the GM counter) detects radioactivity, the flask is shattered by an internal mechanical arrangement, which releases the poison that kills the cat (Figure 1.11). From a quantum perspective, it is implicit that after a while, the cat is simultaneously dead and alive, before making any observation inside the box. However, when one looks into the box, i.e., makes an observation or measurement, he sees that the cat is either alive or dead and not both alive and dead. Hence, it may be interpreted that the act of observation or measurement tends to affect the system itself.

Further, consider another example of an RLC network, as shown in Figure 1.12. Suppose an ammeter is inserted into the loop for the sake of observation or measurement. It may affect the network or system, since the ammeter might have a resistance value. In theory, an ideal ammeter has zero resistance; however, in reality, the ammeter resistance is not zero, which leads to a voltage drop across the ammeter.

The effects of observations are clearly visible in the case of quantum systems, where the motion of a quantum particle can be hindered through observation or measurement made on the particle. In other words, there is an inhibition of transitions between quantum states by frequent measurements. This phenomenon is known as the *quantum Zeno effect*.

26 *Mathematical Modelling of Systems and Analysis*

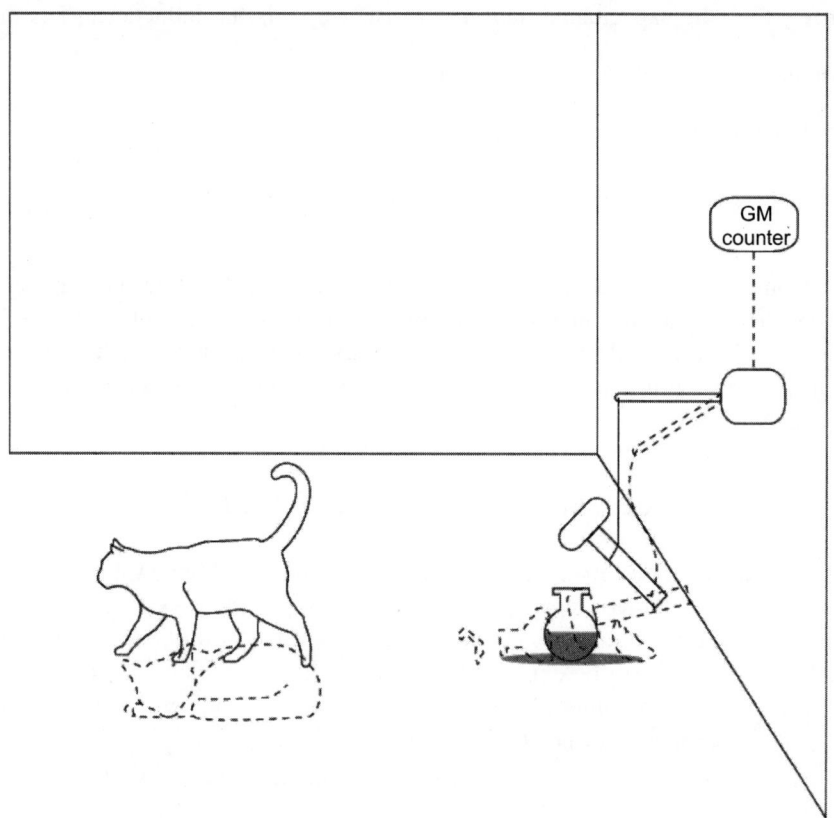

FIGURE 1.11 Schrödinger's cat experiment.

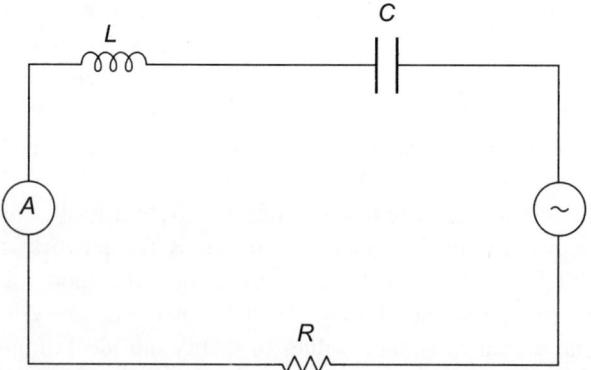

FIGURE 1.12 RLC network with an ammeter for measurement of current in the loop.

EXERCISES

1. Give a few examples for reversible and irreversible processes.
2. A vehicle weighing 2500 kg travels up an inclined plane with a constant speed of 72 km/h. (see the Figure below). Calculate the power required to overcome the frictional force and the force of gravity.

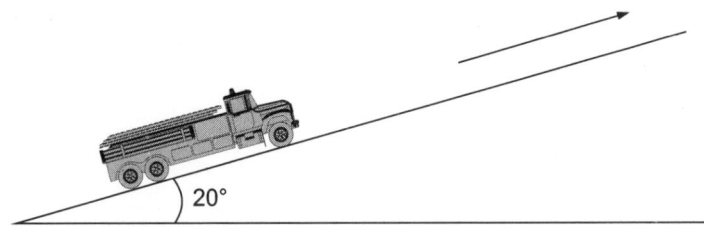

3. A block of weight 300 N is dropped on a spring of stiffness 12 N/mm from a height of 150 mm (see the Figure below). Find the compression of the spring if the block is brought to rest.

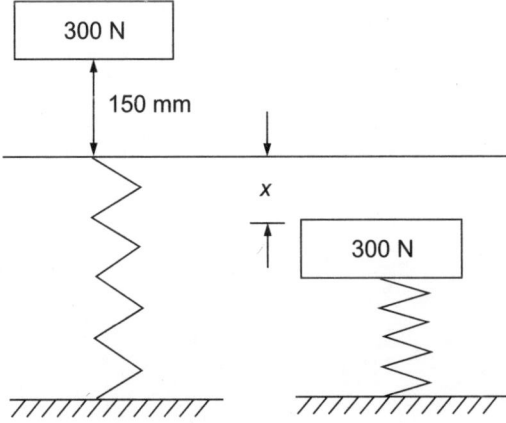

4. Obtain the equation of motion for a particle moving along a straight rigid wire which is rotating on a plane with one end fixed, with angular velocity ω.
5. Explain the mechanism of converting heat into mechanical work.
6. Discuss the impossibility of a perpetual motion machine.
7. Derive the expression for entropy as a function of temperature, pressure and volume using thermodynamic relationships.

8. Mathematically show that entropy does not change upon any reversible change in the state of a closed system.
9. Mathematically show that the entropy of a closed system grows in an irreversible process.
10. Can change in entropy be a criterion of the equilibrium and spontaneity of a process?
11. Find the change in entropy in the reversible isothermal compression of one mole of oxygen from $P_1 = 0.001$ atm to $P_2 = 0.01$ atm.
12. Find out the relationship between Schrödinger's cat experiment and the existence of quantum particles as waves and particles.

CHAPTER 2
Classification and Representations of Systems

Systems can be classified based on their structure, spatial and temporal dependence, inputs and functionality. At the most basic level, systems can be classified into closed and open systems. A *closed system* is a system that does not interact with the environment and is not affected or influenced by the surroundings. An *open system* is a system whose boundaries allow exchange of energy, material and information with the environment.

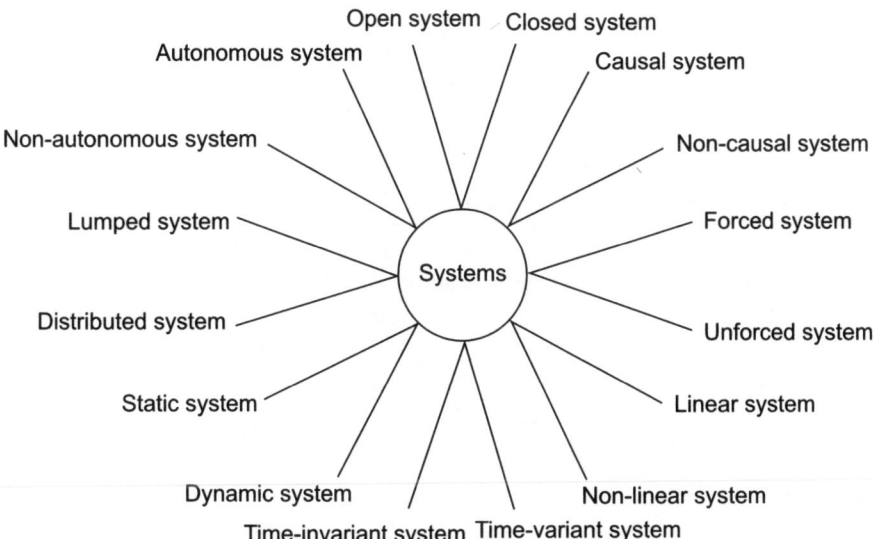

FIGURE 2.1(a) Classification of systems based on structure.

Based on their structure [see Figure 2.1(a)], systems can be classified into static and dynamic systems, lumped and distributed systems, causal and non-causal systems, forced and unforced systems, time-invariant and time-variant systems, autonomous and non-autonomous systems, linear and non-linear systems, hybrid systems, Zeno systems, etc. Further, systems can be classified based on the area of application into physical systems, biological systems, chemical systems, mechanical systems, electrical systems, physiological systems, psychological systems, medical systems, ecological systems, sociological systems, coupled systems such as electromechanical systems, etc. [see Figure 2.1(b)].

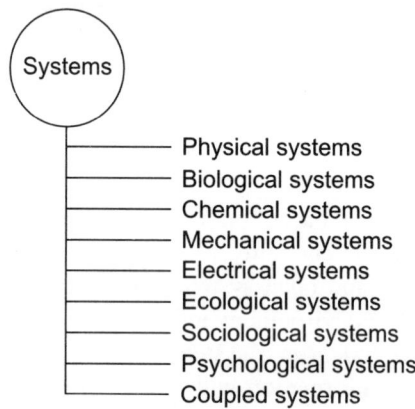

FIGURE 2.1(b) Classification of systems based on the area of application.

Now, in further sections, we will discuss the types of systems based on structure in detail.

2.1 STATIC AND DYNAMIC SYSTEMS

Static systems are those systems whose output at any instant of time depends on the input values at the same instant of time. For example, consider a resistor with resistance R. The equation relating the input current i to output voltage v can be expressed as $v(t) = R.i(t)$. The output voltage at any time instant t depends on the input current at the same tth instant of time.

Consider another example given by the following equation:

$$y(t) = 9x^2(t) + 10x(t)$$

where, y is the output and x is the input. Also, in this case, the output at tth instant of time depends on the input at tth instant. Hence, it is a static system.

Static systems are referred to as *memory-less* or *memory-free systems*, since the outputs and inputs do not depend on their values at previous time instants.

Dynamic systems are those systems whose outputs at the present time instant depend on the previous values of inputs as well as on the previous values of outputs. Such systems are considered as systems having memory and memory elements. The function of a memory element (Figure 2.2) is to retain or remember values over a period of time. For example, consider the dynamical system represented by the following equation:

$$y(t) = x(t) + 5\,x(t-1) + 9\,y(t-1)$$

Here, the output at *t*th instant of time depends on input at the *t*th instant ($x(t)$), input at the previous instant ($x(t-1)$) and also on the output at the previous instant ($y(t-1)$).

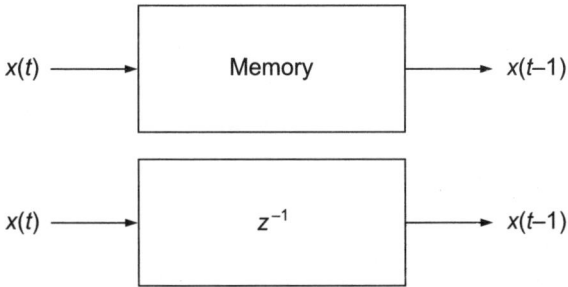

FIGURE 2.2 Representations of a memory element.

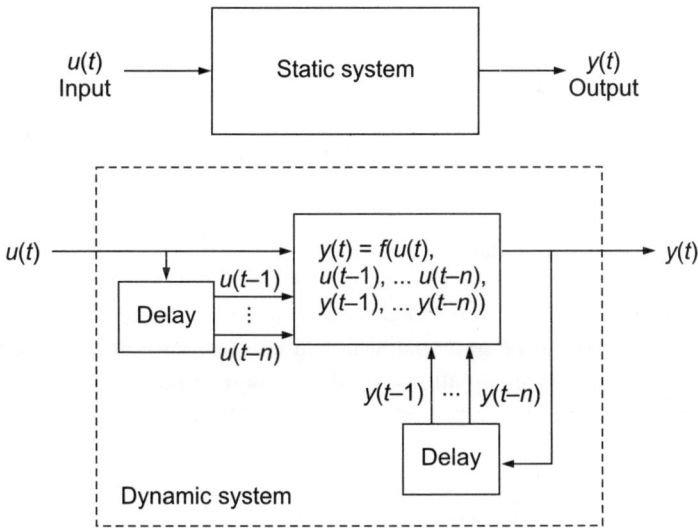

FIGURE 2.3 Representations of static and dynamic systems.

2.2 LUMPED AND DISTRIBUTED SYSTEMS

A *lumped system* is defined as a system whose dependent variables (state variables) are a function of time alone. Such systems are represented using ordinary differential equations (ODEs). For example, consider the system shown in Figure 2.4, described by the equation $M\dfrac{d^2x(t)}{dt^2} + B\dfrac{dx(t)}{dt} + Kx(t) = 0$, where, M, B and K represents the mass, damping coefficient and spring constant, respectively. Here, all the state variables, namely, displacement, velocity and acceleration $\left(x, \dfrac{dx}{\partial t} \text{ and } \dfrac{d^2x}{\partial t^2}\right)$ vary as a function of time only.

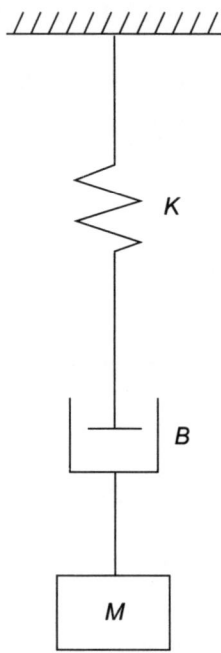

FIGURE 2.4 Mass-spring-dashpot system.

A *distributed system* is defined as a system whose dependent variables are functions of both time and spatial positions. Such systems are represented using partial differential equations (PDEs). For example, consider the one-dimensional heat equation represented as follows:

$$\dfrac{\partial u}{\partial t} = \dfrac{\partial^2 u}{\partial x^2}$$

In this case, the evolution of the state variable u depends both on time t and length x.

2.3 CAUSAL AND NON-CAUSAL SYSTEMS

A *causal system* is defined as a system whose output depends on the present and past inputs and not on the future inputs. For example,

$$y(t) = x(t) + x(t-3)$$

Generally, all real-time systems are causal systems, since in real-time applications, only present and past samples are present.

A *non-causal system* or *anti-causal system* is defined as a system whose present response depends on future values of inputs. For example,

$$y(t) = x(t) + 10\, x(t+1)$$

Since anti-causal systems contain future samples, these systems are practically not realisable.

2.4 FORCED AND UNFORCED SYSTEMS

A *forced system* (Figure 2.5) is a dynamical system whose output evolution depends on forcing function or inputs. An *unforced system* (Figure 2.6) is a system whose output evolution does not depend on any forcing function or inputs.

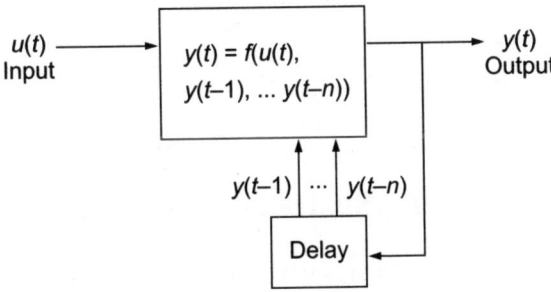

FIGURE 2.5 Representation of a forced system.

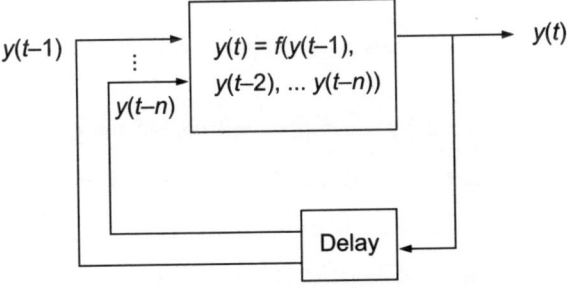

FIGURE 2.6 Representation of an unforced system.

2.5 CONSERVATIVE AND NON-CONSERVATIVE SYSTEMS

In the action of any dynamical system, energy is dissipated usually through some form of friction. However, in some systems, the dissipation of energy is very slow and can be neglected in relatively short interval of time. In such systems, we assume the law of conservation of energy, which states that the sum of the kinetic energy, and potential energy is constant. The systems for which this law holds good are known as *conservative systems*.

For example, consider a mass-spring system, which is a simple conservative system. If x is the displacement of mass M from its equilibrium position and the restoring force exerted on M by the spring is $-kx$, where $k > 0$ is the spring constant or the stiffness of the spring, then the equation of motion is written as

$$M\frac{d^2x}{dt^2} + kx = 0$$

Now, consider the case where a damping force is introduced in the system (i.e., M moves through a resisting medium). The equation of motion for such a system is written as

$$M\frac{d^2x}{dt^2} + B\frac{dx}{dt} + kx = 0$$

where, B is the damping coefficient.

This system is a typical non-conservative system, since there is significant dissipation of energy due to the damping force $\left(-B \cdot \dfrac{dx}{dt}\right)$.

2.6 TIME-INVARIANT AND TIME-VARIANT SYSTEMS

A *time-invariant system* is defined as a system whose input-output characteristics do not change with time. In other words, the input-output characteristics of such systems do not change with time shifting. For example, let us check whether the system represented by the difference equation [Eq. (2.1)]

$$y(t) = x(t) - 5x(t - 2) \qquad (2.1)$$

is time-invariant or not.

Let us delay the input by k samples and denote the output by $y(t, k)$.

$$y(t, k) = x(t - k) - 5\,x(t - 2 - k) \qquad (2.2)$$

Now, replacing t by $t - k$ throughout the given Eq. (2.1), we get

$$y(t - k) = x(t - k) - 5x(t - k - 2) \qquad (2.3)$$

Comparing Eqs. (2.2) and (2.3), we see that $y(t, k) = y(t - k)$. Hence, the given system is time-invariant.

Further, *time-invariant systems* can also be defined as those systems whose system parameters do not change as a function of time. For example, consider a mechanical system described in Figure 2.4, where K is the spring constant, B is the damping coefficient, M is the mass and x is the displacement. The governing equation of the system can be expressed as

$$M\frac{d^2x(t)}{dt^2} + B\frac{dx(t)}{dt} + Kx(t) = 0$$

Since the parameters of the system (M, B, and K) are constants and do not change as a function of time, the system is a time-invariant system.

A *time-variant system* is a system whose input-output characteristics change with time shifting. For example, let us check whether the system represented by the difference equation [Eq. (2.4)]

$$y(t) = x(t) + t\, x(t-1) \qquad (2.4)$$

is time variant or not.

Similar to the previous case, let us delay the input by k samples and denote the output by $y(t, k)$.

$$y(t, k) = x(t-k) + t\, x(t-1-k) \qquad (2.5)$$

Now, replacing t by $t-k$ throughout Eq. (2.4), we get

$$y(t-k) = x(t-k) + (t-k)\, x(t-k-1) \qquad (2.6)$$

Comparing Eqs. (2.5) and (2.6), we see that $y(t, k) \neq y(t-k)$. Hence, the given system is time variant.

A *time-variant system* can also be defined as a system whose parameters change as a function of time. Consider the mass-spring-dashpot system shown in Figure 2.4 and assume that the spring constant K varies as a function of time due to environmental effects and aging of the material of the spring. Now, the governing equation of the system can be expressed as

$$M\frac{d^2x(t)}{dt^2} + B\frac{dx(t)}{dt} + K(t)x(t) = 0$$

Now, since K changes with respect to time, the system is a time-variant system.

2.7 AUTONOMOUS AND NON-AUTONOMOUS SYSTEMS

A dynamic system is said to be an *autonomous system* if its output evolution does not depend explicitly on time. Such a system is of the form $\frac{dx}{dt} = f(x)$.

Whereas, a dynamic system is said to be a *non-autonomous system* if its output evolution depends explicitly on time. A non-autonomous system is of the form $\frac{dx}{dt} = f(t, x)$. Such systems are more complicated, since we need information on both x and t to predict the future state of the system.

2.8 HYBRID SYSTEMS

A *hybrid system* is defined as a dynamic system that exhibits both continuous and discrete dynamic behaviour. Such a system is represented by both continuous differential equations and discrete difference equations.

2.9 ZENO SYSTEMS

A *Zeno system* is a system which performs an infinite number of discrete steps in a finite amount of time.

2.10 LINEAR SYSTEMS

A *linear system* is a system which follows *superposition principle*. The superposition principle states that for all linear systems, the total response at any given time instant caused by two or more inputs is the sum of the outputs which would have been caused by each input individually. In other words, if input u_1 produces response y_1 and input u_2 produces response y_2 then input $(u_1 + u_2)$ produces response $(y_1 + y_2)$. Further, if the input to the linear system is zero, then it produces zero output.

For example, let us check the linearity of the system whose relationship between the input x and the output y is described by the equation $y(t) = t\, x(t)$. When input $x(t)$ is zero, then the output is also zero. Here, the first test for linearity is satisfied. Now, let us check whether the superposition principle is satisfied for this system.

Let $x_1(t)$ and $x_2(t)$ be the two inputs which produces outputs $y_1(t)$ and $y_2(t)$ respectively.

$$y_1(t) = t\, x_1(t) \tag{2.7}$$

$$y_2(t) = t\, x_2(t) \tag{2.8}$$

Now, adding Eqs. (2.7) and (2.8), we get

$$g(t) = y_1(t) + y_2(t) = t\, x_1(t) + t\, x_2(t) = t\, [x_1(t) + x_2(t)]$$

Now, let us add $x_1(t)$ and $x_2(t)$ and apply this input to the system.

$$h(t) = t[x_1(t) + x_2(t)]$$

It is seen that $g(t) = h(t)$. Hence, superposition principle is valid and the given system is linear.

In the context of systems governed by linear differential equations, the principle of superposition states that if x_1 is a solution of $\dfrac{dx}{dt} = Ax + f_1$ and x_2 is a solution of $\dfrac{dx}{dt} = Ax + f_2$, then $x = x_1 + x_2$ is a solution of $\dfrac{dx}{dt} = Ax + f_1 + f_2$.

This statement can be proved easily. Since x_1 is a solution of $\dfrac{dx}{dt} = Ax + f_1$, we get

$$\frac{dx_1}{dt} = Ax_1 + f_1 \qquad (2.9)$$

and, since x_2 is a solution of $\dfrac{dx}{dt} = Ax + f_2$, we get

$$\frac{dx_2}{dt} = Ax_2 + f_2 \qquad (2.10)$$

Adding Eqs. (2.9) and (2.10) gives

$$\frac{d(x_1 + x_2)}{dt} = A(x_1 + x_2) + f_1 + f_2$$

This implies that $x_1 + x_2$ is a solution of $\dfrac{dx}{dt} = Ax + f_1 + f_2$.

For example, consider $\dfrac{dx}{dt} = Ax + f_1$, where $A = \begin{pmatrix} 1 & -1 \\ -2 & 0 \end{pmatrix}$ and $f_1 = \begin{pmatrix} t \\ 1 \end{pmatrix}$. Consider $\dfrac{dx}{dt} = Ax + f_2$, where $f_2 = \begin{pmatrix} t^2 - t + 1 \\ 4t \end{pmatrix}$. It can be checked that $x_1 = \begin{pmatrix} e^{2t} + e^{-t} \\ -e^{2t} + 2e^{-t} + t \end{pmatrix}$ is a solution of $\dfrac{dx}{dt} = Ax + f_1$ and $x_2 = \begin{pmatrix} e^{2t} + e^{-t} + t \\ -e^{2t} + 2e^{-t} + t^2 \end{pmatrix}$ is a solution of $\dfrac{dx}{dt} = Ax + f_2$. Now, $x = x_1 + x_2 = \begin{pmatrix} 2e^{2t} + 2e^{-t} + t \\ -2e^{2t} + 4e^{-t} + t + t^2 \end{pmatrix}$ is a solution of $\dfrac{dx}{dt} = Ax + f_1 + f_2$.

2.11 NON-LINEAR SYSTEMS

A *non-linear system* is defined as a system which does not follow superposition principle. Further, non-linear systems are characterised as follows:

1. The input-output relationship is non-linear. The static relationships are non-linear.
2. The system comprises both linear and non-linear elements.
3. Superposition principle is not valid.
4. Stability is complex.

5. Complicated relationships are involved.
6. Non-linear systems are modelled using non-linear differential equations and analytical solution is difficult to obtain.

2.11.1 PARASITIC NON-LINEARITY AND VOLUNTARY NON-LINEARITY

A *parasitic non-linearity* is an unwanted non-linearity which creeps into the system without the knowledge of the user. Such nonlinearities degrade the performance of the system, for example, friction.

A *voluntary non-linearity* is a non-linearity that is deliberately introduced into the system by the user for the purpose of analysis. For example, in order to measure flow rate in a pipe using a differential pressure transmitter, the user introduces a square root extractor for flow measurement, since the flow rate is proportional to square root of the pressure difference.

2.11.2 TYPICAL PHYSICAL NON-LINEARITIES

Some common examples of physical non-linearities are saturation, hysteresis, dead zone, friction and relay.

Saturation: In saturation non-linearity, the output is proportional to the input only for a limited range of input values. When the input exceeds this range, the output tends to become unchanged or saturated (Figure 2.7). The saturation non-linearity is often exhibited by amplifiers and control valves.

Hysteresis: A hysteresis non-linearity is a type of non-linearity which is commonly exhibited by electromagnetic systems. If the input is increased from zero to fullscale and then reduced from fullscale to zero, two different sets of outputs are produced. This phenomenon is known as *hysteresis* (Figure 2.8).

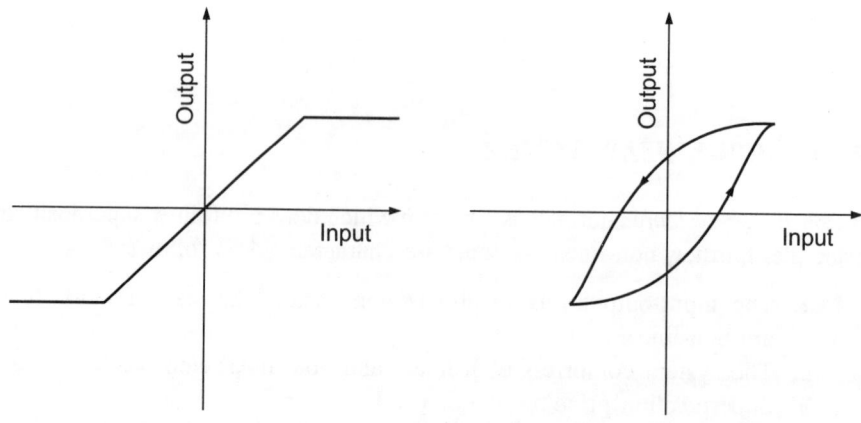

FIGURE 2.7 Saturation.   **FIGURE 2.8** Hysteresis.

Friction: Friction is an opposing force acting on the relative motion of the body (Figure 2.9). The friction non-linearity is often exhibited by mechanical systems.

Dead zone: The dead zone non-linearity refers to a condition in which output becomes zero for a certain range of input values (Figure 2.10). This type of non-linearity is exhibited by various electrical devices like motors and actuators.

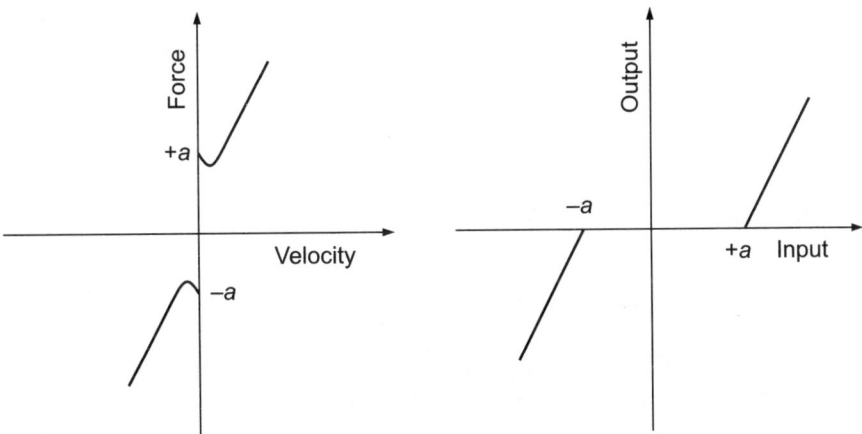

FIGURE 2.9 Friction.   **FIGURE 2.10** Dead zone.

Relay: A relay is a type of non-linear element that switches on if the input exceeds zero and switches off if the input goes below zero (Figure 2.11). Such elements are often used in ON-OFF control systems. Figures 2.12 and 2.13 show relay elements which exhibit properties of hysteresis and dead zone type of nonlinearities, respectively.

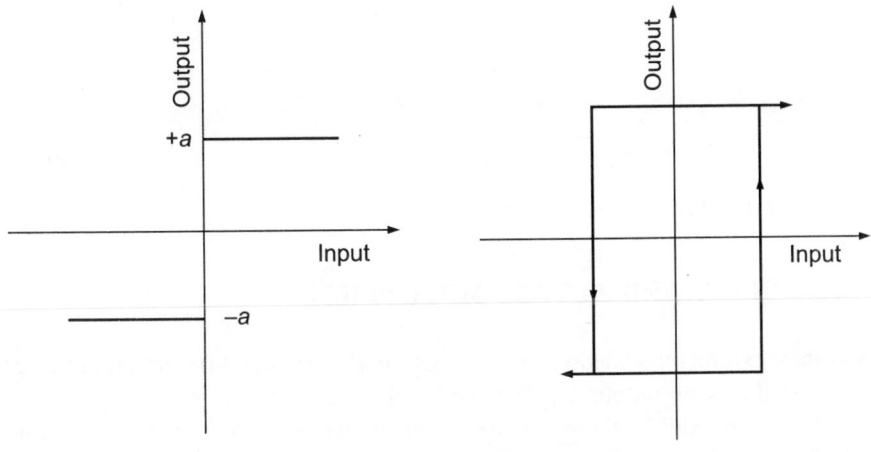

FIGURE 2.11 Relay.   **FIGURE 2.12** Relay with hysteresis.

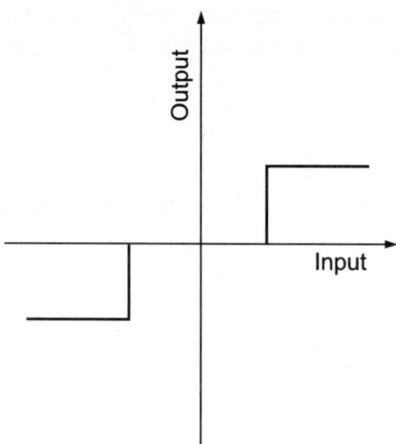

FIGURE 2.13 Relay with dead zone.

2.11.3 General Non-linearity

The general non-linearity is the type of non-linearity in which the relation between the input and output is non-linear (Figure 2.14).

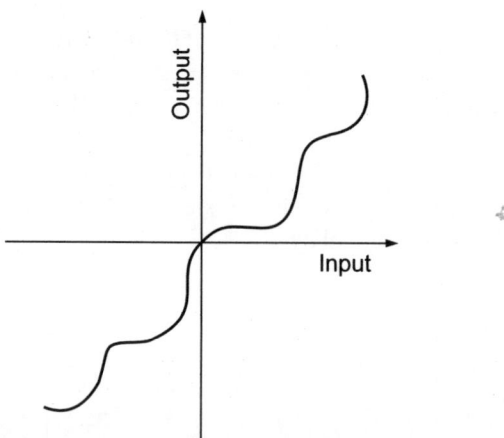

FIGURE 2.14 General non-linearity.

2.12 INPUT AND OUTPUT MULTIPLICITY

A system exhibits the phenomenon of *input multiplicity* if different input values can give the same output, as described in Figure 2.15.

A system exhibits the phenomenon of *output multiplicity* if a single input value can give rise to different output values, as described in Figure 2.16.

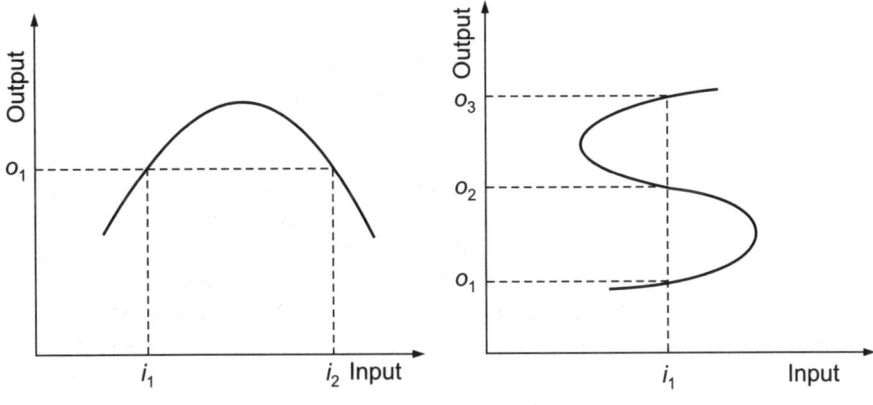

FIGURE 2.15 Input multiplicity.   **FIGURE 2.16** Output multiplicity.

2.13 LINEARISATION

Linearisation is the technique of finding the linear approximation of a non-linear function. Static non-linear relationships can be linearised using the 'line of best fit' method and the piecewise linearisation technique. In the *line of best fit* approach, a straight line passing through the maximum number of points of the input-output graph is drawn and the equation of the straight line given by $y = mx + c$ represents the linear approximation of the non-linear relation (Figure 2.17). Further, in the piecewise linearisation approach, a non-linear curve is segmented into smaller units and each unit is separately linearised (Figure 2.18). This type of linearisation yields a discontinuous relation between the input and output, however, it gives a better representation of the non-linear curve than the line of best fit approach.

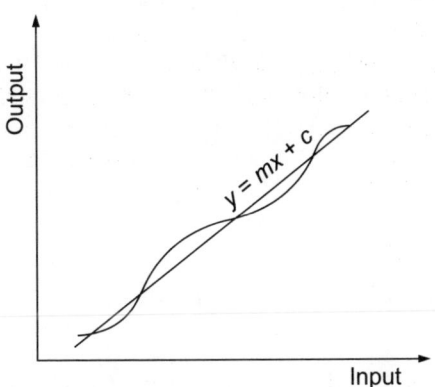

FIGURE 2.17 The 'line of best fit' approach for linearisation.

A non-linear dynamic system can be linearised using either the Taylor's series approximation or the Jacobian approach. Linearisation of non-linear

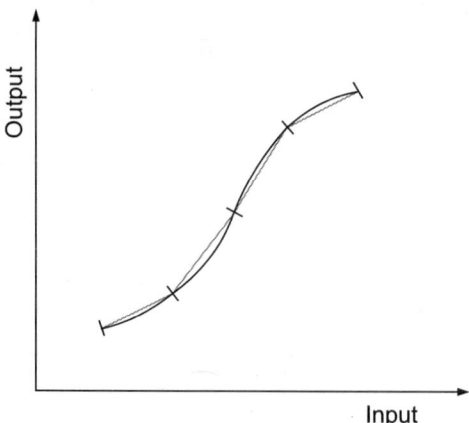

FIGURE 2.18 The piecewise linearisation approach.

systems leads to severe approximation errors; however, it is useful in some cases to linearise a non-linear system around an operating point for the purpose of analysis and for the design of controllers for controlling the actual system.

2.13.1 Taylor's Series Method for Linearisation

Consider a non-linear system represented by the following simultaneous differential equations:

$$\frac{dx}{dt} = f(x, y)$$

$$\frac{dy}{dt} = g(x, y)$$

where, $f(x, y)$ and $g(x, y)$ are nonlinear functions. To linearise the system around the operating points $(\bar{x}, \bar{y})$, the Taylor's series approximation can be used as follows:

$$f(x, y) \approx f(\bar{x}, \bar{y}) + \frac{\partial f}{\partial x}\bigg|_{(\bar{x}, \bar{y})} (x - \bar{x}) + \frac{\partial f}{\partial y}\bigg|_{(\bar{x}, \bar{y})} (y - \bar{y})$$

$$g(x, y) \approx g(\bar{x}, \bar{y}) + \frac{\partial g}{\partial x}\bigg|_{(\bar{x}, \bar{y})} (x - \bar{x}) + \frac{\partial g}{\partial y}\bigg|_{(\bar{x}, \bar{y})} (y - \bar{y})$$

For example, let us linearise the system governed by the non-linear equation $\frac{dx}{dt} = \sqrt{x}$, using Taylor's series method.

Let $f(x) = \sqrt{x}$ and applying Taylor's series approximation, we get

$$\sqrt{x} \approx \sqrt{\bar{x}} + \frac{1}{2\sqrt{\bar{x}}} (x - \bar{x})$$

Hence, the linearised system is

$$\frac{dx}{dt} = \sqrt{\bar{x}} + \frac{1}{2\sqrt{\bar{x}}} (x - \bar{x})$$

Consider another example where we would like to linearise the system of simultaneous differential equations given as follows:

$$\frac{dx}{dt} = 6y - 2x^2$$

$$\frac{dy}{dt} = 4y - xy$$

Using Taylor's series method, around the operating points (2, 2).
The non-linear terms present in this equation are x^2 and xy.
Let us consider the following:

$$f(x, y) = 6y - 2x^2$$

$$g(x, y) = 4y - xy$$

Now, applying Taylor's series expansion, we get

$$f(x, y) \approx 4 - 8(x - 2) + 6(y - 2)$$

$$g(x, y) \approx 4 - 2(x - 2) + 2(y - 2)$$

Hence, the linearised system is

$$\frac{dx}{dt} = -8x + 6y + 8$$

$$\frac{dy}{dt} = -2x + 2y + 4$$

2.13.2 The Jacobian Approach for Linearisation

Consider a non-linear system represented by

$$\frac{dx}{dt} = f(x, y)$$

$$\frac{dy}{dt} = g(x, y)$$

where, $f(x, y)$ and $g(x, y)$ are non-linear functions. To linearise the system around the operating points $(\bar{x}, \bar{y})$, the Jacobian approach can be used as follows:

$$\begin{pmatrix} \dfrac{dx}{dt} \\ \dfrac{dy}{dt} \end{pmatrix} = \begin{pmatrix} \dfrac{\partial f}{\partial x}_{(\bar{x},\bar{y})} & \dfrac{\partial f}{\partial y}_{(\bar{x},\bar{y})} \\ \dfrac{\partial g}{\partial x}_{(\bar{x},\bar{y})} & \dfrac{\partial g}{\partial y}_{(\bar{x},\bar{y})} \end{pmatrix} \begin{pmatrix} x - \bar{x} \\ y - \bar{y} \end{pmatrix}$$

The Jacobian approach directly yields the matrix vector representation of the set of non-linear differential equations. The Jacobian matrix

$$A = \begin{pmatrix} \dfrac{\partial f}{\partial x}_{(\bar{x},\bar{y})} & \dfrac{\partial f}{\partial y}_{(\bar{x},\bar{y})} \\ \dfrac{\partial g}{\partial x}_{(\bar{x},\bar{y})} & \dfrac{\partial g}{\partial y}_{(\bar{x},\bar{y})} \end{pmatrix}$$ is known as the *system matrix* and can be directly

used for stability analysis of the system.

Consider an example where we would like to linearise the system given by:

$$\frac{dx}{dt} = 10(y - x)$$

$$\frac{dy}{dt} = x(28 - z) - y$$

$$\frac{dz}{dt} = xy - \frac{8}{3}z$$

using the Jacobian approach.

$$f(x, y, z) = 10(y - x)$$

Now, let $g(x, y, z) = x(28 - z) - y$

$$h(x, y, z) = xy - \frac{8}{3}z$$

Using Jacobian approach,
$$\begin{pmatrix} \dfrac{dx}{dt} \\ \dfrac{dy}{dt} \\ \dfrac{dz}{dt} \end{pmatrix} = \begin{pmatrix} \dfrac{\partial f}{\partial x}_{(\bar{x},\bar{y},\bar{z})} & \dfrac{\partial f}{\partial y}_{(\bar{x},\bar{y},\bar{z})} & \dfrac{\partial f}{\partial z}_{(\bar{x},\bar{y},\bar{z})} \\ \dfrac{\partial g}{\partial x}_{(\bar{x},\bar{y},\bar{z})} & \dfrac{\partial g}{\partial y}_{(\bar{x},\bar{y},\bar{z})} & \dfrac{\partial g}{\partial z}_{(\bar{x},\bar{y},\bar{z})} \\ \dfrac{\partial h}{\partial x}_{(\bar{x},\bar{y},\bar{z})} & \dfrac{\partial h}{\partial y}_{(\bar{x},\bar{y},\bar{z})} & \dfrac{\partial h}{\partial z}_{(\bar{x},\bar{y},\bar{z})} \end{pmatrix} \begin{pmatrix} x - \bar{x} \\ y - \bar{y} \\ z - \bar{z} \end{pmatrix}$$

Hence, the linearised system around the operating points $(\bar{x}, \bar{y}, \bar{z})$ is given by

$$\begin{pmatrix} \dfrac{dx}{dt} \\ \dfrac{dy}{dt} \\ \dfrac{dz}{dt} \end{pmatrix} = \begin{pmatrix} -10 & 10 & 0 \\ 28-\bar{z} & -1 & -\bar{x} \\ \bar{y} & \bar{x} & -\dfrac{8}{3} \end{pmatrix} \begin{pmatrix} x-\bar{x} \\ y-\bar{y} \\ z-\bar{z} \end{pmatrix}$$

2.14 THE DESCRIBING FUNCTION

Consider a non-linear input-output relation given by $y(t) = f(x(t))$, where f is a non-linear function. Now, consider a sinusoidal input $x(t) = X \sin \omega t$. For this input, the output $y(t)$ may not be a sinusoidal function, but a periodic function, and can be expressed as a Fourier series.

$$y(t) = A_0 + \sum_{n=1}^{\infty} [A_n \cos(n\omega t) + B_n \sin(n\omega t)]$$

$$= A_0 + \sum_{n=1}^{\infty} Y_n \sin(n\omega t + \phi_n)$$

$$Y_n = \sqrt{A_n^2 + B_n^2}$$

$$\phi_n = \arctg \frac{A_n}{B_n}$$

For symmetric non-linearity, $A_0 = 0$ and the harmonic of $y(t)$ could be neglected.

$$y(t) = Y_1 \sin(\omega t + \phi_1)$$

Now, we can describe the non-linear components by the frequency response. This is known as the *describing function*. The describing function $N(X)$ of the non-linear element is the complex ratio of the fundamental component of the output $y(t)$ and the sinusoidal input $x(t)$. For,

$$x(t) = X \sin \omega t$$
$$y(t) = A_1 \cos \omega t + B_1 \sin \omega t$$
$$= Y_1 \sin(\omega t + \phi_1)$$

$$N(X) = \frac{Y_1 e^{j\phi_1}}{X}$$

$$Y_1 = \sqrt{A_1^2 + B_1^2}$$

$$\phi_1 = \operatorname{arctg} \frac{A_1}{B_1}$$

$$A_1 = \frac{1}{\pi} \int_0^{2\pi} y(t) \cos \omega t \, d(\omega t)$$

$$B_1 = \frac{1}{\pi} \int_0^{2\pi} y(t) \sin \omega t \, d(\omega t)$$

The describing function is actually the linearised frequency response (harmonic linearisation).

2.14.1 Steps to Obtain the Describing Function of a Non-linear Element

1. Input a sinusoidal signal $x(t)$ to the non-linear elements

$$x(t) = X \sin \omega t$$

2. Solve $y(t)$ to obtain its fundamental component.
3. Calculate the describing function $N(X)$ using the following formulas:

$$y(t) = Y_1 \sin(\omega t + \phi_1)$$

$$\left. \begin{array}{l} A_1 = \dfrac{1}{\pi} \displaystyle\int_0^{2\pi} y(t) \cos \omega t \, d(\omega t) \\[6pt] B_1 = \dfrac{1}{\pi} \displaystyle\int_0^{2\pi} y(t) \sin \omega t \, d(\omega t) \\[6pt] Y_1 = \sqrt{A_1^2 + B_1^2} \\[6pt] \phi_1 = \operatorname{arctg} \dfrac{A_1}{B_1} \end{array} \right\}$$

$$N(X) = \frac{Y_1 e^{j\phi_1}}{X}$$

For example let us determine the describing function of the device whose mathematical description is given by

$$y(t) = \frac{1}{2} x(t) + \frac{1}{4} x^3(t)$$

Considering $x = X \sin \omega t$, and passing it through the given function, we get

$$y(t) = \frac{1}{2} X \sin \omega t + \frac{1}{4} X^3 \sin^3 \omega t$$

$$= \frac{1}{2} X \sin \omega t + \frac{1}{4} X^3 \left[\frac{3 \sin \omega t - \sin 3\omega t}{4} \right]$$

$$= \frac{1}{2} X \sin \omega t + \frac{3}{16} X^3 \sin \omega t - \frac{1}{16} X^3 \sin 3\omega t$$

$$= \left(\frac{1}{2} X + \frac{3}{16} X^3 \right) \sin \omega t - \frac{1}{16} X^3 \sin \omega t$$

Neglecting harmonics, we get

$$y_1(t) = \left(\frac{1}{2} X + \frac{3}{16} X^3 \right) \sin \omega t \text{ and } \phi_1 = 0$$

Hence the describing function is

$$N(X) = \frac{Y_1}{X} = \frac{\frac{1}{2} X + \frac{3}{16} X^3}{X} = \frac{1}{2} + \frac{3}{16} X^2$$

EXERCISES

1. Differentiate between lumped and distributed parameter models with examples.
2. Explain parasitic and voluntary non-linearity with examples.
3. What is the necessity for linearisation of non-linear systems?
4. Linearise the following system using Jacobean approach:

    $$\dot{x} = -y - z$$
    $$\dot{y} = x + \alpha y$$
    $$\dot{z} = \beta + xz - \gamma z$$

5. Linearise the following system using Taylor's series method:

    $$\frac{dx}{dt} = x^{1/3}$$

6. Briefly explain 'input multiplicity' and 'output multiplicity'.
7. Explain the characteristics of a non-linear system.
8. Derive the describing function of a relay.

CHAPTER 3
Mathematical Modelling of Systems

Some scientists and mathematicians consider mathematics as the language of the Gods and that God has created the entire universe using a set of mathematical equations. Mathematics plays an important role in placing new hypothesis, proving certain hypothesis and disproving existing ones. For example, consider the dichotomy principle. Two thousand years ago, a Greek philosopher and mathematician named Zeno raised a certain hypothesis. Consider an object moving in a straight line from point A to B. Before covering the whole distance and arriving at point B, an object must first cover half that distance. Then, it must cover half of the remaining distance, and so on. Hence, the object must cover infinite fractional distances before reaching B, which leads to the conclusion that the destination point B will never be reached. Further, the motion cannot even begin because the moving object, before moving half the distance, must move a quarter of the distance, etc. *ad infinitum* so that the object cannot even begin to move. This leads to the conclusion that motion as well as distance must be an illusion. This is known as the *dichotomy paradox* or the *Zeno paradox,* and the rationale behind the paradox is even now considered true by several mathematicians and scientists.

For the purpose of analysing systems, two different models can be developed—the first one being the physical model and the second being the mathematical model. The physical model of the system is an actual or a real model, where the model is a small-scale physical representation of the system with full functionality or reduced functionality. This type of model is like a prototype of the system and is useful for analysing the features, functionality and stability of the system in real environmental conditions. For example, consider the wind tunnel experiment on an aircraft model, where the physical model of the aircraft is a geometrically equivalent structure of the actual aircraft, however with reduced size and preserving the geometrical ratios. This model is fixed inside the wind tunnel and a stream of air is passed through the tunnel. The effect of the wind speed on the aircraft model and other associated dynamics is analysed using several sensors for measuring different forces acting on the aircraft. Another example of a physical model is the polyacrylamide gel, which is a tissue-mimicking material. The polyacrylamide gel can be used to mimic the properties of real animal tissues by changing the concentration of the acrylamide monomer, and hence, can be used for experimental analysis.

A mathematical model of a system is a description of a real-world system using mathematical equations, concepts and language. Mathematical representations of real world systems are highly useful for analysing the system, predicting the future states of the system, assessing the stability of the system, and also, for the development of suitable controllers for controlling the real-world systems. Mathematical models are used to simplify the application of theory. It is in the context of modelling that engineers and mathematicians can work together with biologists, physicists, doctors, psychologists and professionals from various other fields to aid in scientific discovery.

If a mathematical model is developed using only theoretical concepts, then such a model may fail when it is made to perform real-time computations. On the other hand, some models may be developed entirely based on measurement data. These models may lack the power of generalisation or the power to adapt when subjected to a new set of data. Hence, measured data as well as abstract concepts are required to develop a realistic mathematical model. In some cases, measurement data is used to estimate the unknown parameters of a model whose structure is well known or measurements may be needed to validate a model developed using theoretical concepts.

3.1 THE REGRESSION LINE AND CORRELATION VALUE

The regression line is one of the most basic and essentially important methods for modelling the relationship between two variables. Consider two data variables x and y. Suppose if there are n subjects indexed by $i = 1, 2, ..., n$, then each data variable stores a value for each subject for a particular experiment. For example, consider that x_i is the moisture content and y_i is the mould growth rate. Before computing the regression line, it is necessary to compute the following descriptive statistics of the data set:

1. The average of x and y

$$\bar{x} = \frac{1}{n} \sum_{i=1}^{n} x_i$$

$$\bar{y} = \frac{1}{n} \sum_{i=1}^{n} y_i$$

2. The variance of x and y

$$\text{Var } x = \frac{1}{n} \sum_{i=1}^{n} (x_i - \bar{x})^2$$

$$\text{Var } y = \frac{1}{n} \sum_{i=1}^{n} (y_i - \bar{y})^2$$

3. The standard deviation (SD) of x and y

   SD of x is $s_x = \sqrt{\text{Var } x}$

   SD of y is $s_y = \sqrt{\text{Var } y}$

4. The correlation coefficient (r)

$$r = \frac{1}{n} \sum_{i=1}^{n} \left(\frac{x_i - \bar{x}}{s_x} \cdot \frac{y_i - \bar{y}}{s_y} \right)$$

where, $-1 \le r \le 1$.

The regression line of y on x passes through the point of averages $(\bar{x}, \bar{y})$, whose slope is $r\frac{s_y}{s_x}$. The regression of y on x, given by the line $y = mx + c$, can be used to predict y from x. The error or residual for subject i is $e_i = y_i - c - mx_i$. The mean square error (MSE) is given by

$$\text{MSE} = \frac{1}{n} \sum_{i=1}^{n} e_i^2$$

For the regression line, the MSE is equal to $(1 - r^2)$ Var y. Figure 3.1 shows a typical linear regression line which relates the dependent variable y to the independent variable x.

The correlation value indicates how strongly y is related to x. In other words, the correlation value is a quantitative measure of the relationship between x and y. If the sign of the correlation coefficient is positive, then the regression line slopes up. If the sign is negative, the line slopes down. A negative correlation indicates that an increase in x results in a decrease in y. Table 3.1 presents a common way of interpreting the results of correlation analysis.

TABLE 3.1 Interpretation of the Correlation Value

Correlation value	Relation between x and y
1	Perfect positive correlation
0.7 to 0.99	Very strong positive correlation
0.4 to 0.69	Strong positive correlation
0.3 to 0.39	Moderate positive correlation

(Contd.)

Mathematical Modelling of Systems **51**

TABLE 3.1 Interpretation of the Correlation Value (Contd.)

Correlation value	Relation between x and y
0.2 to 0.29	Weak positive correlation
0.01 to 0.19	No or negligible positive correlation
0	No relationship exists
–0.01 to –0.19	No or negligible negative correlation
–0.2 to –0.29	Weak negative correlation
–0.3 to –0.39	Moderate negative correlation
–0.4 to –0.69	Strong negative correlation
–0.7 to –0.99	Very strong negative correlation
–1	Perfect negative correlation

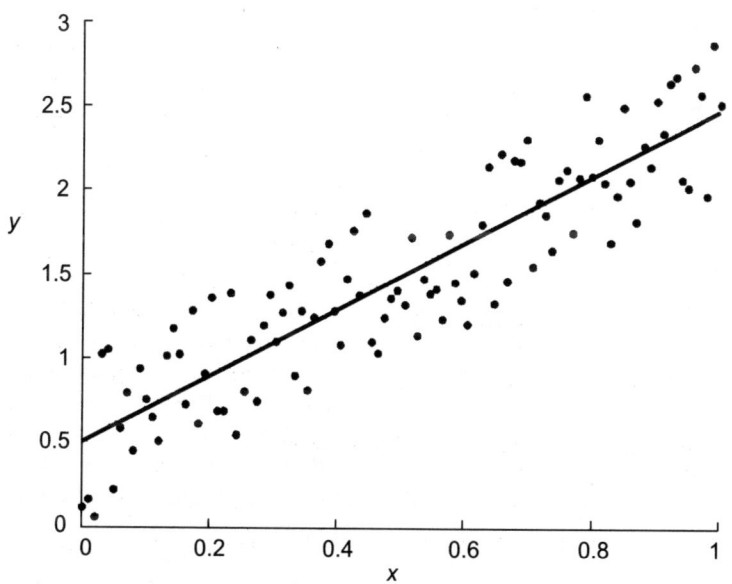

FIGURE 3.1 The regression line.

Now, consider that a tensile test is conducted on a particular material and the test results are given in Table 3.2.

TABLE 3.2 A typical result of a tensile test

Load (kg)	Length of elongation (cm)
0	2
1	2.1
2	2.22
3	2.31
4	2.43
5	2.52

The regression model for the above experiment is given by

$$y = mx + c$$

where, y is the elongation length and x is the applied load. In the above equation, m and c are constants that are dependent on the mechanical properties of the material considered and are estimated from the data. The computed regression line for this case is shown in Figure 3.2.

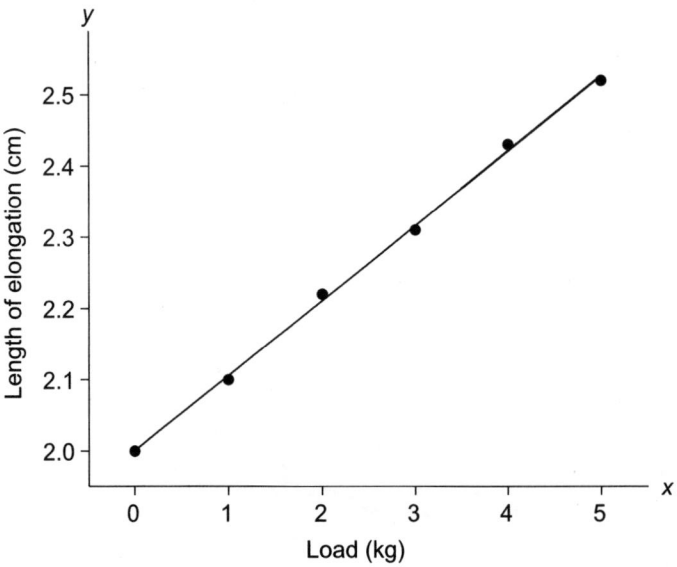

FIGURE 3.2 Computed regression line for data in Table 3.2.

Now, the equation can be used to predict the elongation of the material for loads greater than 5 kg assuming that the material characteristics are linear elastic.

3.2 MULTIPLE REGRESSION

Consider that Y is an $n \times 1$ vector, which is a dependent or response variable. Let Y_i be the ith component of Y. Consider that X is an $n \times p$ matrix of observable variables. X is called the design matrix. Assume that $n > p$ and the columns of X are linearly independent (i.e., the design matrix has full rank or the rank of X is p).

Consider that X is related to Y by the following regression equation:

$$Y = X\beta + \varepsilon$$

where, β is a $p \times 1$ vector of parameters, which needs to be estimated from data, and ε is an $n \times 1$ random vector known as the random error or disturbance term and is generally not observed. Now, the regression equation can be broken into n equations, i.e., one equation for each observation.

$$Y_i = X_i \beta + \varepsilon_i$$

where, X_i is the ith row of X and ε_i is the ith element of ε.

Now, using the observed values of X and Y, β can be estimated using ordinary least squares (OLS) estimator.

$$\hat{\beta} = (X^T X)^{-1} X^T Y$$

where, X^T denotes the transpose of X and $\hat{\beta}$ denotes the estimated values of β.

3.3 POLYNOMIAL REGRESSION

Polynomial regression can be used to map the relationship between the dependent and the independent variable when the relationship is not linear and is curvilinear. The general structure of the polynomial regression model is given by

$$y = a_0 + a_1 x + a_2 x^2 + a_3 x^3 + \ldots + a_n x^n + \varepsilon$$

where, y and x are the dependent and independent variables, respectively, and $[a_0, a_1, a_2, a_3, \ldots, a_n]$ are the polynomial coefficients, which can be estimated using ordinary least squares (OLS) when measurements of y and x are available. n is the order of the polynomial. If $n = 1$, then the model is actually a linear regression model. $n = 2$ and $n = 3$ correspond to quadratic and cubic polynomials, respectively. As a general rule, the accuracy of the fit increases with the increase in the order of the polynomial and the order of the polynomial can be $n-1$ if n observations are available. Figure 3.3 shows a typical cubic polynomial relating variables y and x.

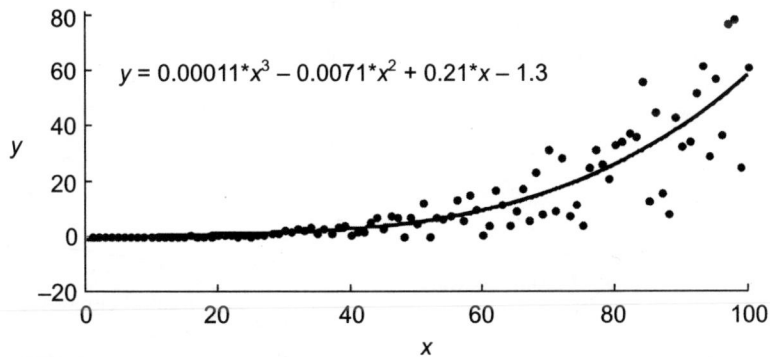

FIGURE 3.3 A cubic polynomial regression.

Polynomial regression is highly useful for the development of empirical models because of the following advantages:

1. Polynomials can effectively capture the trend of the data.
2. Polynomial models can be integrated and differentiated.
3. With sufficient data points, the estimation of polynomial coefficients is quite easy.

However, in some cases, the use of polynomials is not advised because of the following disadvantages:

1. Higher order polynomial models exhibit oscillations near the end points of the data interval.
2. Lower order polynomial models fail to capture the trend of data with large variations.

3.4 MODELLING APPLICATIONS IN MARKETING

Most of the mathematical modelling methodologies find applications in a variety of domains. Consider the supply and demand for a product A. The supply and demand curves show the relationship between quality and price. The supply curve shows the quantity of the product which is bought to the market at different prices. The demand curve shows the total quantity of the product that the consumers will buy at various prices (Figures 3.4 and 3.5). The supply and demand curves are represented as follows if the relationship between price and quality is linear

$$\text{Supply:} \quad Q = a_0 + a_1 P$$
$$\text{Demand:} \quad Q = b_0 + b_1 P$$

A quadratic relationship between quality and demand is given as follows:

$$\text{Supply:} \quad Q = a_0 + a_1 P + a_2 P^2$$
$$\text{Demand:} \quad Q = b_0 + b_1 P + b_2 P^2$$

where, a_0, a_1, a_2, b_0, b_1 and b_2 are parameters.

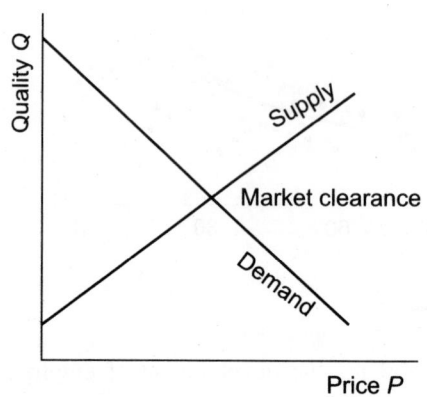

FIGURE 3.4

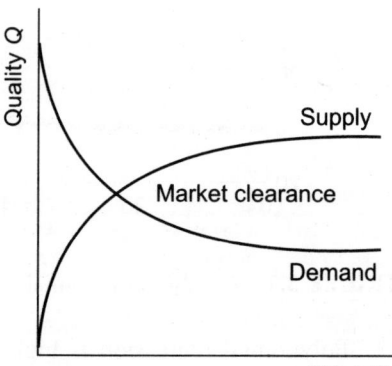

FIGURE 3.5

The free market price is determined by the crossing point of the two curves (Freedman, 2009). At the free market price, market clearance is achieved, i.e., the supply is equal to demand. The supply is affected by the cost of production. Let W be the labour rate and H be the price of raw materials required for production of product A. The demand is affected by prices of complements, which go along with the product, and substitutes, which can replace the product. For example, if the product is butter, then the complement is usually bread and the substitute is jam. Let T be the price of the complements and M be the price of substitutes. Now, framing the model which relates the quality and price, we obtain

$$\text{Supply:} \quad Q = a_0 + a_1 P + a_2 W + a_3 H + \delta_t$$
$$\text{Demand:} \quad Q = b_0 + b_1 P + b_2 T + b_3 M + \varepsilon_t$$

where, a_0, a_1, a_2, a_3, b_0, b_1, b_2 and b_3 are parameters and δ_t and ε_t are random disturbance terms.

3.5 SPLINE INTERPOLATION

To overcome the disadvantages of the polynomial regression, we go for a method known as *spline interpolation* or *spline functions*. In polynomial regression, a function f is approximated over the entire interval using one polynomial of higher degree. However, in the method of splines, we do partition of the entire interval into many sub-intervals and approximate the function f using lower degree polynomials. Such polynomials are called spline functions.

Let $[a, b]$ be a finite interval and $a = x_0 < x_1 < x_2 < x_3 \ldots < x_n = b$ be the partition points. A spline function $S(x)$ of degree m has the following properties.

1. $S(x)$ is $m-1$ times continuously differentiable on $[a, b]$
2. On each sub-interval $[x_{i-1}, x_i]$, $i = 1, 2, \ldots, n$, $S(x)$ is a polynomial of degree m.
3. $S(x_i) = f_i$.

In cubic spline interpolation, $S(x)$, $S'(x)$ and $S''(x)$ are continuous on $[a, b]$ and $S(x)$ is a cubic polynomial $S_i(x)$ on each sub-interval $[x_{i-1}, x_i]$, $i = 1, 2, \ldots, n$. Let $S(x_i) = f_i$, $i = 1, 2, \ldots, n$ and $h_i = x_i - x_{i-1}$, $i = 1, 2, \ldots, n$.

$$d_i = \frac{f_i - f_{i-1}}{h_i}, \quad i = 1, 2, \ldots, n \text{ and } t = \frac{x - x_{i-1}}{h_i} \text{ for } x \in [x_{i-1}, x_i].$$

Since $S_i(x)$ is a cubic polynomial in the interval $[x_{i-1}, x_i]$ satisfying the conditions $S_i(x_{i-1}) = f_{i-1}$ and $S_i(x_i) = f_i$, then we have

$$S_i(x) = (1-t)f_{i-1} + tf_i + h_i t(1-t)[(1-t)A_i + tB_i] \qquad (3.1)$$

where, A_i and B_i need to be determined.

Differentiating Eq. (3.1), we get

$$S_i'(x) = d_i + (1-2t)[(1-t)A_i + tB_i] + (t-t^2)(B_i - A_i) \quad (3.2)$$

Let

$$S_i'(x_i) = k_i \quad \text{and} \quad S_i'(x_{i-1}) = k_{i-1}$$

Using $t = 0$ and $t = 1$, from Eq. (3.2), we get

$$A_i = k_{i-1} - d_i$$
$$B_i = d_i - k_i$$

Substituting A_i and B_i in Eq. (3.1), we get

$$S_i(x) = (1-t)f_{i-1} + tf_i + h_i t(1-t)[(k_{i-1} - d_i)(1-t) - (k_i - d_i)t] \quad (3.3)$$

From Eq. (3.3), it is seen that $S_{i+1}'(x_i) = k_i$. Hence, $S'(x)$ is continuous. Differentiating Eq. (3.3) twice, we get

$$S_i''(x) = \frac{2}{h_i}[3d_i - 2k_{i-1} - k_i - 3t(2d_i - k_{i-1} - k_i)]$$

Hence, $$S_i''(x_i) = \frac{2}{h_i}[k_{i-1} + 2k_i - 3d_i] \quad (3.4)$$

and $$S_{i+1}''(x_i) = \frac{2}{h_{i+1}}[3d_{i+1} - 2k_i - k_{i+1}] \quad (3.5)$$

For $S''(x)$ to be continuous, we must have $S_i''(x_i) = S_{i+1}''(x_i)$. Hence, equating Eqs. (3.4) and (3.5), we get

$$\frac{k_{i-1}}{h_i} + 2\left(\frac{1}{h_i} + \frac{1}{h_{i+1}}\right)k_i + \frac{k_{i+1}}{h_{i+1}} = 3\left(\frac{d_i}{h_i} + \frac{d_{i+1}}{h_{i+1}}\right), \quad i = 1, 2, \ldots, n-1. \quad (3.6)$$

Equation (3.6) represents $n - 1$ equations in $n + 1$ unknowns $k_0, k_1, \ldots, k_n$

We assume $$S''(x_0) = S''(x_n) = 0 \quad (3.7)$$

The obtained spline is called a *natural spline*.

From Eq. (3.4), (3.5), and (3.7), we get

$$\left.\begin{array}{l} 2k_0 + k_1 = 3d_i \\ k_{n-1} + 2k_n = 3d_n \end{array}\right\} \quad (3.8)$$

By placing Eq. (3.6) in between Eq. of (3.8), we get

$$\begin{bmatrix} 2 & 1 & 0 & 0 & \cdots & 0 \\ \dfrac{1}{h_1} & 2\left(\dfrac{1}{h_1}+\dfrac{1}{h_2}\right) & \dfrac{1}{h_2} & 0 & \cdots & 0 \\ 0 & \dfrac{1}{h_2} & 2\left(\dfrac{1}{h_2}+\dfrac{1}{h_3}\right) & \dfrac{1}{h_3} & \cdots & \cdots \\ \cdots & \cdots & \cdots & \cdots & \cdots & \cdots \\ \cdots & \cdots & \cdots & \cdots & \cdots & \cdots \\ 0 & 0 & 0 & \cdots & 1 & 2 \end{bmatrix} \begin{bmatrix} k_0 \\ k_1 \\ k_2 \\ \vdots \\ \vdots \\ k_n \end{bmatrix} = 3 \begin{bmatrix} d_1 \\ \dfrac{d_1}{h_1}+\dfrac{d_2}{h_2} \\ \dfrac{d_2}{h_2}+\dfrac{d_3}{h_3} \\ \vdots \\ \vdots \\ d_n \end{bmatrix} \quad (3.9)$$

These equations can be solved to get the unknowns k_0, k_1, ..., k_n. By substituting these values in Eq. (3.3), we can find $S_i(x)$ for each of the sub intervals.

If $h_1 = h_2 = ... = h_n$, then Eq. (3.9) becomes

$$\begin{bmatrix} 2 & 1 & 0 & \cdots & 0 \\ 1 & 4 & 1 & \cdots & 0 \\ 0 & 1 & 4 & \cdots & 0 \\ \vdots & \vdots & \vdots & \vdots & \vdots \\ \vdots & \vdots & \vdots & \vdots & \vdots \\ 0 & 0 & 0 & 1 & 2 \end{bmatrix} \begin{bmatrix} k_0 \\ k_1 \\ k_2 \\ \vdots \\ \vdots \\ k_n \end{bmatrix} = 3 \begin{bmatrix} d_1 \\ d_1+d_2 \\ d_2+d_3 \\ \vdots \\ \vdots \\ d_n \end{bmatrix}$$

EXAMPLE 3.1 Find the cubic spline interpolation for the data given in the following table.

x	f
1	1
2	0
3	1
4	0
5	1

Solution: In this case $h_1 = h_2 = ... = h_n = 1$, $d_1 = -1$, $d_2 = 1$, $d_3 = -1$, $d_4 = 1$.

$$\begin{bmatrix} 2 & 1 & 0 & 0 & 0 \\ 1 & 4 & 1 & 0 & 0 \\ 0 & 1 & 4 & 1 & 0 \\ 0 & 0 & 1 & 4 & 1 \\ 0 & 0 & 0 & 1 & 2 \end{bmatrix} \begin{bmatrix} k_0 \\ k_1 \\ k_2 \\ k_3 \\ k_4 \end{bmatrix} = 3 \begin{bmatrix} -1 \\ 0 \\ 0 \\ 0 \\ 1 \end{bmatrix} = \begin{bmatrix} -3 \\ 0 \\ 0 \\ 0 \\ 3 \end{bmatrix}$$

On solving, we get

$$k_0 = \frac{-12}{7}, \; k_1 = \frac{3}{7}, \; k_2 = 0, \; k_3 = \frac{-3}{7}, \; k_4 = \frac{12}{7}$$

Hence, In [1, 2], $S_1(x) = 1 - \dfrac{12t}{7} + \dfrac{5t^3}{7}, \; t = x - 1$

In [2, 3], $S_2(x) = \dfrac{3t}{7} + \dfrac{15t^2}{7} - \dfrac{11t^3}{7}, \; t = x - 2$

In [3, 4], $S_3(x) = 1 - \dfrac{18t^2}{7} + \dfrac{11t^3}{7}, \; t = x - 3$

In [4, 5], $S_4(x) = -\dfrac{3t}{7} + \dfrac{15t^2}{7} - \dfrac{5t^3}{7}, \; t = x - 4$

The following figure shows the cubic spline interpolation for data given in the table:

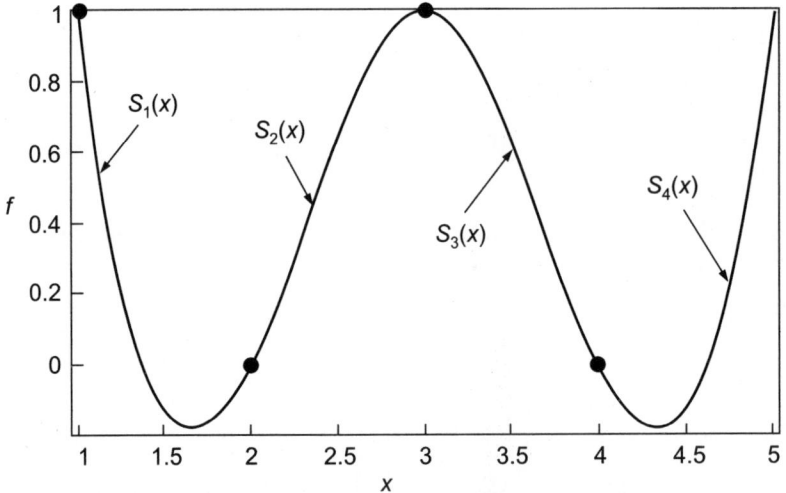

3.6 MODELLING OF DOSE CONCENTRATION

As a non-linear case, consider a mathematical equation to model the variation in the concentration of a drug in blood. The drug concentration in blood usually decreases as a function of time since the drug is eliminated from the body (Thomas and Finney, 1992). The equation to represent this phenomenon would be

$$R = C_0 e^{-kt}$$

where, R is the residual concentration, C_0 is the concentration administered, k is the elimination constant and t is time in hours.

Usually, doses are repeated to maintain the desired concentration in blood (Figure 3.6). The model for the effects of repeated doses can be given as

$$R_n = C_0 e^{-kt_0} + C_0 e^{-2kt_0} + C_0 e^{-3kt_0} + \ldots + C_0 e^{-nkt_0}$$

where, C_0 is the dose concentration, R_n is the residual concentration before the $(n + 1)$th dose and t_0 is the time between the doses.

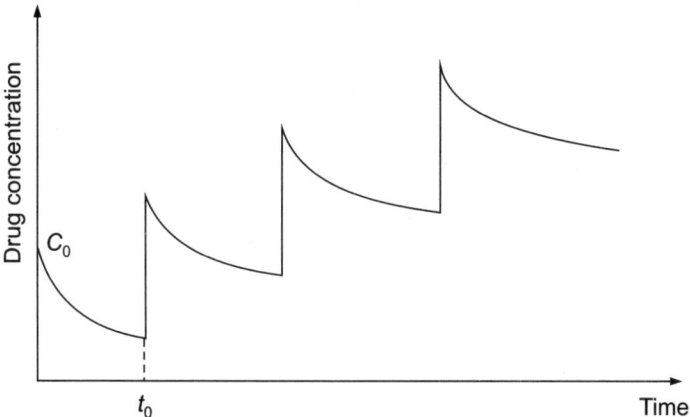

FIGURE 3.6 Drug concentrations in blood due to repeated doses.

3.7 WORKING WITH COMPLEX NUMBERS

In some cases, it becomes necessary to work with complex numbers and functions involving complex variables. For example, the dielectric spectrum of a material is often expressed as a complex variable (varying with angular velocity ω) whose real part is the dielectric permittivity and the imaginary part is the conductivity or dielectric loss. Also, in mechanics, the mechanical modulus is expressed as a complex variable whose real part is the elastic modulus and the imaginary part is the viscous modulus. Hence, in some cases, mathematical models involving complex variables need to be developed for efficient analysis of systems under investigation.

Complex numbers z are defined as ordered pairs $z = (x, y)$, where x and y are real and imaginary parts of z, respectively.

$$\text{Re}(z) = x \text{ and } \text{Im}(z) = y$$

Now, let S be a set of complex numbers. A function f defined on S is a mapping that assigns a complex number w to each z in S.

$$w = f(z)$$

Suppose, $w = u + iv$ is the value of a function f at $z = x + iy$. Then,

$$u + iv = f(x + iy)$$

Here, u and v depend on both x and y.

For example, if $f(z) = z^2$, then

$$f(z) = (x+iy)^2 = x^2 - y^2 + 2ixy$$

Hence, $u = x^2 - y^2$ and $v = 2xy$.

Thus, the given function of the complex variable z can be written in terms of two real-valued functions of variables x and y.

$$f(z) = u(x, y) + iv(x, y)$$

A complex-valued function f is continuous at a point z_0 if $\lim_{z \to z_0} f(z)$ exists, $f(z_0)$ exists and $\lim_{z \to z_0} f(z) = f(z_0)$.

The derivative of f at z_0 given as $f'(z_0)$ is defined by

$$f'(z_0) = \lim_{z \to z_0} \frac{f(z) - f(z_0)}{z - z_0}$$

if this limit exists. The function f is said to be differentiable at z_0 if its derivative at z_0 exists.

Let c be a complex constant and f be a function whose derivative exists at a point z. Then,

$$\frac{dc}{dz} = 0$$

$$\frac{dz}{dz} = 1$$

$$\frac{d[cf(z)]}{dz} = cf'(z)$$

$$\frac{dz^n}{dz} = nz^{n-1}$$

if n is a positive integer.

If the derivatives of two functions f and g exist at a point z, then

$$\frac{d}{dz}[f(z) + g(z)] = f'(z) + g'(z)$$

$$\frac{d}{dz}[f(z)g(z)] = f(z)g'(z) + f'(z)g(z)$$

$$\frac{d}{dz}\left[\frac{f(z)}{g(z)}\right] = \frac{g(z)f'(z) - f(z)g'(z)}{[g(z)]^2}$$

The Debye model is an excellent modelling example with complex valued function. This model characterises the dielectric spectrum of materials at different frequencies. The electrical relaxation of a material is described by the real and imaginary parts of the complex relative permittivity ε as a function of angular frequency ω (Figure 3.7). Here,

$$\varepsilon = \varepsilon' + i\varepsilon''$$

where, ε' is the dielectric permittivity and ε'' is the dielectric loss. The Debye model describes the electrical relaxation in a given material, using parameters, namely, the dielectric increment $\varepsilon_s - \varepsilon_\infty = \Delta\varepsilon$ (where, ε_∞ is the permittivity at field frequencies when $\omega\tau \gg 1$, and ε_s is the permittivity at $\omega\tau \ll 1$ and $i^2 = -1$, which is the magnitude of the dispersion, and the relaxation time τ.

The first step involved in the development of the model is to find an expression of the form $\varepsilon(\omega) = \varepsilon_\infty + f(\omega)$ which reduces to $f(0) = \varepsilon_s - \varepsilon_\infty$ for $\omega = 0$.

The polarisation of a dielectric decreases after removal of an electric field as a function of time. The polarisation decays exponentially with the time constant τ, when the steady field is switched off, and hence, the characteristic relaxation time of the dipole is given by the following differential equation:

$$\dot{P}(t) = -AP(t)$$

where, $A = 1/\tau$ is the proportionality constant. The solution of the differential equation is:

$$P(t) = P_0 e^{-t/\tau}, \quad P_0 = P(0)$$

Taking Fourier transform to obtain the frequency response,

$$f(\omega) = \int_0^\infty P(t) e^{-i\omega t} dt$$

$$= \frac{P_0}{i\omega + \dfrac{1}{\tau}}$$

Using the condition $f(0) = \varepsilon_s - \varepsilon_\infty$ for $\omega = 0$, the Debye equation is obtained as

$$\varepsilon(\omega) = \varepsilon'(\omega) + i\varepsilon''(\omega) = \varepsilon_\infty + \frac{\varepsilon_s - \varepsilon_\infty}{i\omega\tau + 1}$$

In the case of biological materials such as a soft tissue, the material is more complex due to its structure and composition. In such materials, the broadening of the dispersion is seen and is empirically accounted for by using a distribution parameter α. This introduction of the distribution parameter gives rise to the Cole-Cole model, which is more realistic than the Debye model when complex biological materials are involved.

The Debye model is given by:

$$\varepsilon(\omega) = \varepsilon_\infty + \frac{\Delta\varepsilon}{1 + (i\omega\tau)^{(1-\alpha)}}$$

62 Mathematical Modelling of Systems and Analysis

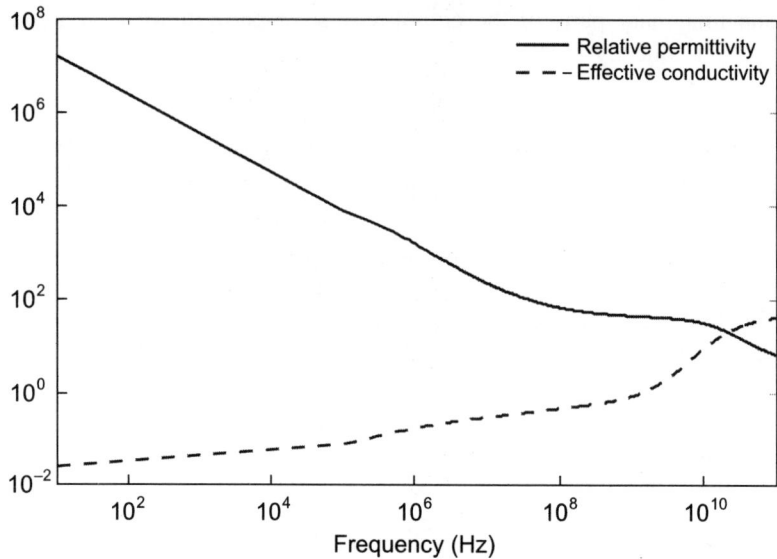

FIGURE 3.7 Variation of permittivity and conductivity of a typical biological soft tissue shown as a function of frequency.

where, α is a measure of the broadening of the dispersion. The dielectric spectrum of biological materials can be appropriately described in terms of multiple Cole-Cole dispersion.

$$\varepsilon(\omega) = \varepsilon_\infty + \sum_n \frac{\Delta\varepsilon_n}{1 + (i\omega\tau_n)^{(1-\alpha_n)}} + \frac{\sigma_i}{i\omega\varepsilon_0}$$

where, $\dfrac{\sigma_i}{i\omega\varepsilon_0}$ is the conductivity term, σ_i is the static ionic conductivity and ε_0 is the permittivity of free space.

3.8 MATHEMATICAL MODELLING USING DIMENSIONAL ANALYSIS

Dimensional analysis serves as an excellent tool for building simple and realistic models from hypothesis.

For example, let us assume that force is a function of mass (m) and acceleration (a).

$$F = f(m, a) \propto m^b a^c$$

We now find the values of the powers (b, c) by writing the equation in terms of the fundamental dimensions, using Table 1.1.

$$MLT^{-2} = M^b (LT^{-2})^c = M^b L^c T^{-2c}$$

By comparing, we obtain the following equations:
$$1 = b,\ 1 = c,\ -2 = -2c$$
Hence, we conclude that $F = m^1 a^1$, which is Newton's second law.

Consider another problem where we would like to find an expression for the force of attraction between two celestial bodies. Let m_1 be the mass of the first body and m_2 be the mass of the second body, and let r be the distance between the two bodies. For this problem, also consider the gravitational constant G. Now, assuming that the force of attraction is a function of m_1, m_2, r, and G, we write

$$F = f(m_1, m_2, r, G) = m_1^a m_2^b r^c G^d$$

Using Table 1.1, we get

$$MLT^{-2} = M^a M^b L^c (M^{-1} L^3 T^{-2})^d$$
$$= M^a M^b M^{-d} L^c L^{3d} T^{-2d}$$
$$= M^{(a+b-d)} L^{(c+3d)} T^{-2d}$$

Equating the powers, we get
$$1 = a + b - d$$
$$1 = c + 3d$$
$$-2 = -2d$$

Hence, we get $d = 1$, $c = -2$ and $a + b = 2$. However, by symmetry, a must be equal to b, and hence, $a = b = 1$. Therefore, the required equation is

$$F = m_1^1 m_2^1 r^{-2} G^1$$
$$= \frac{m_1 m_2 G}{r^2}$$

which is the Newton's law of gravitation.

Consider a third case where we would like to obtain an equation relating the applied stress to the resultant strain of a material.

$$\text{Stress} = f(\text{Strain}) \propto (\text{Strain})^a$$

In dimensional terms,

$$ML^{-1}T^{-2} = (M^0 L^0 T^0)^a$$

We can see that the equation is not satisfied, and hence, we can consider that a dimensional constant, which depends on the type of the material under investigation, must be introduced into the relation. Let λ be the constant. Now, we can write

$$\text{Stress} = \lambda\ \text{Strain} \qquad (3.10)$$

Hence, λ must have the same unit ($ML^{-1}T^{-2}$) of stress. Equation (3.10) is known as Hooke's Law, and λ is known as the Young's modulus of the material.

EXAMPLE 3.2 Find the relationship between the kinetic energy of a gas molecule, Boltzmann constant k and temperature T.

Solution: The Boltzmann constant has the dimension $ML^2T^{-2}\theta^{-1}$. Now, using the arbitrary relation,

$$\text{Kinetic energy (KE)} = \frac{1}{2}mv^2 = f(k, T) \propto k^a T^b$$

In dimensional terms,

$$ML^2T^{-2}\theta^0 = (ML^2T^{-2}\theta^{-1})^a \theta^b$$

Equating the powers,

$$1 = a$$
$$b - a = 0$$
$$b = a = 1$$

Hence, we get

$$\frac{1}{2}mv^2 \propto kT$$

or

$$\frac{1}{2}mv^2 = ckT$$

where, c is a dimensionless constant. Using experimental data, the value of the proportionality constant can be obtained. The equation relating the kinetic energy of a gas molecule to the Boltzmann constant and the temperature is

$$\frac{1}{2}mv^2 = \frac{3}{2}kT$$

3.9 BUCKINGHAM'S π THEOREM

Suppose u is a physical quantity that needs to be expressed as a function of n measurable physical quantities (variables and parameters), $[W_1, W_2, ..., W_n]$, i.e.,

$$u = f(W_1, W_2, ..., W_n) \tag{3.11}$$

The physical quantities $(u, W_1, W_2, ..., W_n)$ have m fundamental dimensions $Q_1, Q_2, ..., Q_m$. In reality, any quantity can be expressed using the fundamental dimensions $Q_1 = M$, $Q_2 = L$, $Q_3 = T$, and $Q_4 = \theta$.

Let Z denotes any of the physical quantity $(u, W_1, W_2, ..., W_n)$. Then, the dimension of Z denoted by $[Z]$ is expressed as a product of powers of the fundamental dimensions $[Z] = Q_1^{\alpha_1} Q_2^{\alpha_2} ... Q_m^{\alpha_m}$, where $(\alpha_1, \alpha_2, ..., \alpha_m)$ are the dimensional exponents of Z and are real numbers. In vector form, the dimensional exponents are expressed by $\alpha = (\alpha_1, \alpha_2, ..., \alpha_m)^t$.

$[Z] = Q_1^{\alpha_1} Q_2^{\alpha_2} \ldots Q_m^{\alpha_m}$ is dimensionless if $(\alpha_1, \alpha_2, \ldots, \alpha_m) = (0, 0, \ldots, 0)$.

Let $b_i = (b_{1i}, b_{2i}, \ldots, b_{mi})^t$ be the dimension vector of W_i, $i = 1, 2, \ldots, n$. Let $B = (b_{ij})$ be the matrix of order $m \times n$ called the dimension matrix of the given problem. It can be observed that Eq. (3.11) can be expressed in terms of dimensionless quantities. The number of dimensionless quantities $k + 1 = n + 1 - r(B)$. Here, $r(B)$ denotes the rank of the matrix B.

Now, let $X^i = (x_{1i}, x_{2i}, \ldots, x_{ni})^t$, $i = 1, 2, \ldots, k$ represents the k linearly independent solutions X of the system $BX = 0$. Let $a = (a_1, a_2, \ldots, a_m)^t$ be the dimension vector of u and $y = (y_1, y_2, \ldots, y_n)^t$ represents a solution of the system $By = -a$. Then, Eq. (3.11) can be written as

$$\pi = g(\pi_1, \pi_2, \ldots, \pi_k)$$

where, $\pi, \pi_1, \pi_2, \ldots, \pi_k$ are dimensionless quantities

$$\pi = u W_1^{y_1} W_2^{y_2} \ldots W_n^{y_n}$$

$$\pi_i = W_1^{x_{1i}} W_2^{x_{2i}} \ldots W_n^{x_{ni}}, \quad i = 1, 2, \ldots, k$$

and g is an unknown function. Hence, Eq. (3.11) becomes

$$u = W_1^{-y_1} W_2^{-y_2} \ldots W_n^{-y_n} g(\pi_1, \pi_2, \ldots, \pi_k)$$

The Buckingham's π theorem states that any meaningful relation $f(W_1, W_2, \ldots, W_n)$ consisting of physical quantities is equivalent to a relation of the form $g(\pi_1, \pi_2, \ldots, \pi_k)$ involving a set of independent dimensionless combinations.

3.10 ASSUMPTIONS AND APPROXIMATIONS

In order to develop a mathematical model of a system, it is sometimes useful to make some engineering approximations. The following approximations are usually adopted for the development of a model:

1. Neglecting small variations and effects
2. Assuming that the environment and surroundings do not affect the system dynamics
3. Using lumped approximations to distributed characteristics
4. Assuming linear relationships between variables
5. Assuming that the parameters of the system do not change with time
6. Neglecting uncertainty and noise

These assumptions normally reduce the number of differential equations or the order of the equations of the developed model. Using lumped approximations to distributed characteristics leads to the development of ordinary differential equations rather than partial differential equations. Further, assuming that the parameters of the system do not change with time leads to the development of time invariant models rather than time-variant models. Finally, neglection of noise and uncertainty helps in avoiding statistical treatment to the models.

3.11 MODELLING APPROACHES

There are three different types of approaches for the development of a dynamic mathematical model of a system, namely, the white box, the black box and the grey box approaches (Figures 3.8 and 3.9).

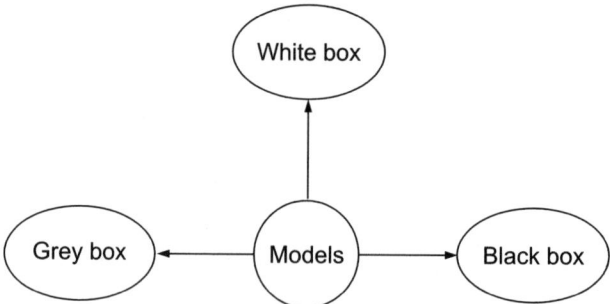

FIGURE 3.8 Modelling approaches.

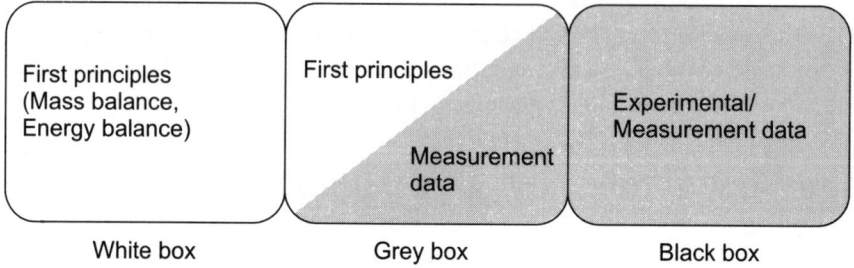

FIGURE 3.9 Representation of white, grey and black box modelling approaches.

White box models

The *white box models* are those models which are fully derived by the first principles. Such models are developed using mass balance or energy balance laws. All the equations and parameters of the white box model are obtained directly from theoretical principles. These models do not depend on data and the parameters of the model are not arbitrary constants and the parameters have a direct relation to the dynamics of the actual system.

The advantage of such models is that they give a very close representation of the actual systems. However, the development of such models is highly complex and requires complete knowledge about the system. Further, the solutions of mathematical models based on first principles are computationally demanding to obtain.

Black box models

The *black box models* are those types of models which are developed solely using experimental data. In this approach, the real-time system is excited using a suitable input and the output from the system is collected. Further, the input-output data obtained from the experiment is utilised to develop the model.

In the black box approach, both the model structure and parameters of the model are determined from the experimental data. The parameters of the black box model have no direct relation to the physical laws governing the actual system. Further, no prior knowledge on the dynamics of the system is required for the development of black box model.

Grey box models

They *grey box models* are a combination of both black and white box models. The knowledge of the first principles or balance laws together with the experimental data is utilised for the development of grey box models. In this case, the model structure is determined from the first principles and the parameters of the model are estimated using the measurement data.

> ***Occam's Razor:*** Occam's Razor or Ockham's razor also known as the law of parsimony can serve as a useful principle for the development of mathematical models as well as for their selection. The statement of the principle can be written as
>
> "Among several competing hypotheses, the one with the fewest assumptions should be selected".
>
> Applying this principle to the development of a mathematical model, we can state that the developed model must have a minimum number of assumptions. If several mathematical models are available for a system under investigation, then the model involving the least number of assumptions must be selected for analysis.
>
> In general, an increase in the number of assumptions will tend to increase the possibility of error, and hence, will reduce the accuracy of the developed model.

Figure 3.10 shows the general flow chart for the development of a mathematical model. A static system can be modelled using linear and non-linear algebraic equations and dynamical systems are modelled using ordinary differential equations and partial differential equations. Figure 3.11 shows the classification of various mathematical models and their applicability in modelling real-world problems. Algebraic models and the models based on ordinary differential equations and partial differential equations are further classified into linear and non-linear models. The solutions to linear models are obtained using direct methods and analytical methods, whereas, the solutions to non-linear models are obtained using indirect and numerical methods.

68 Mathematical Modelling of Systems and Analysis

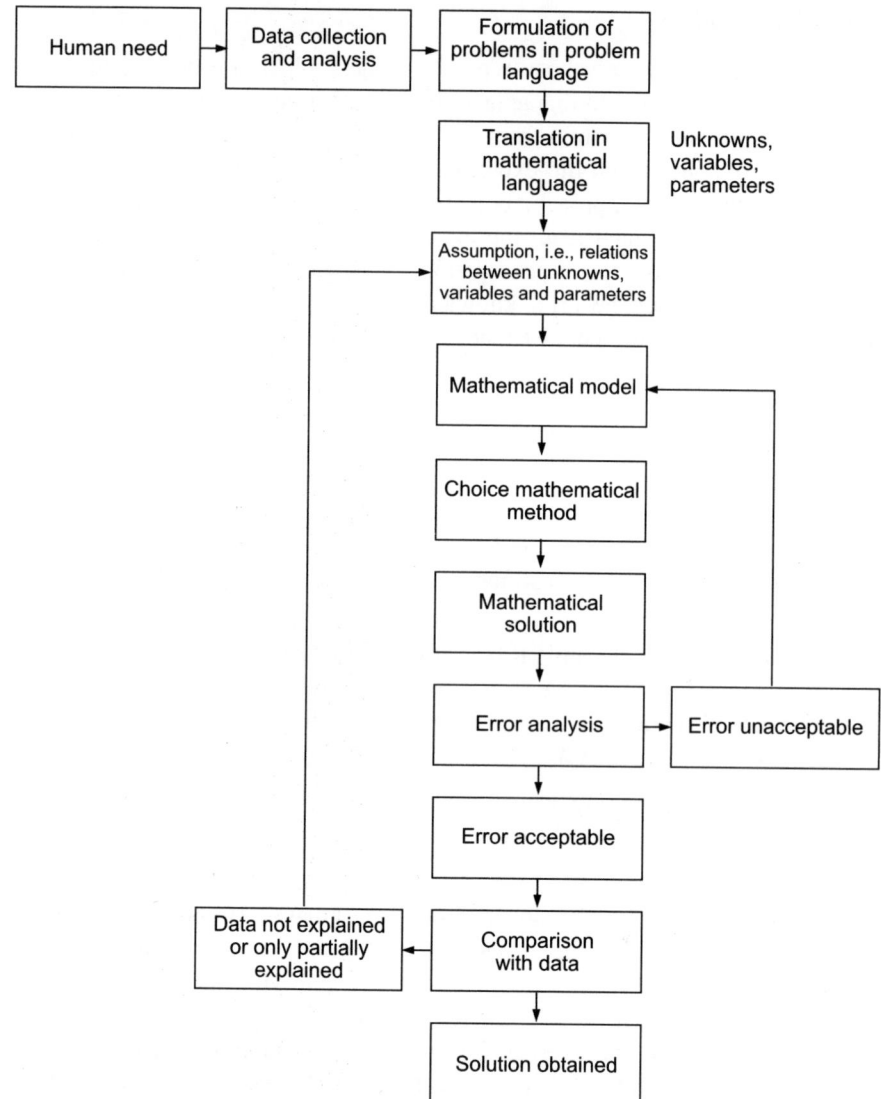

FIGURE 3.10 Flow chart for developing a mathematical model of a real-world system.

The three major applications of a mathematical model are for predicting the future states of the system, for analysing the stability of the system and for designing suitable controllers for controlling the actual system under consideration. Given the initial states of the system, dynamic models can be used for accurately predicting the future states of the system. Such models are often known as *predictive models* and are useful for predicting weather, financial market values, etc. Hence, the development of such models has gained widespread importance in the fields of science, engineering and management.

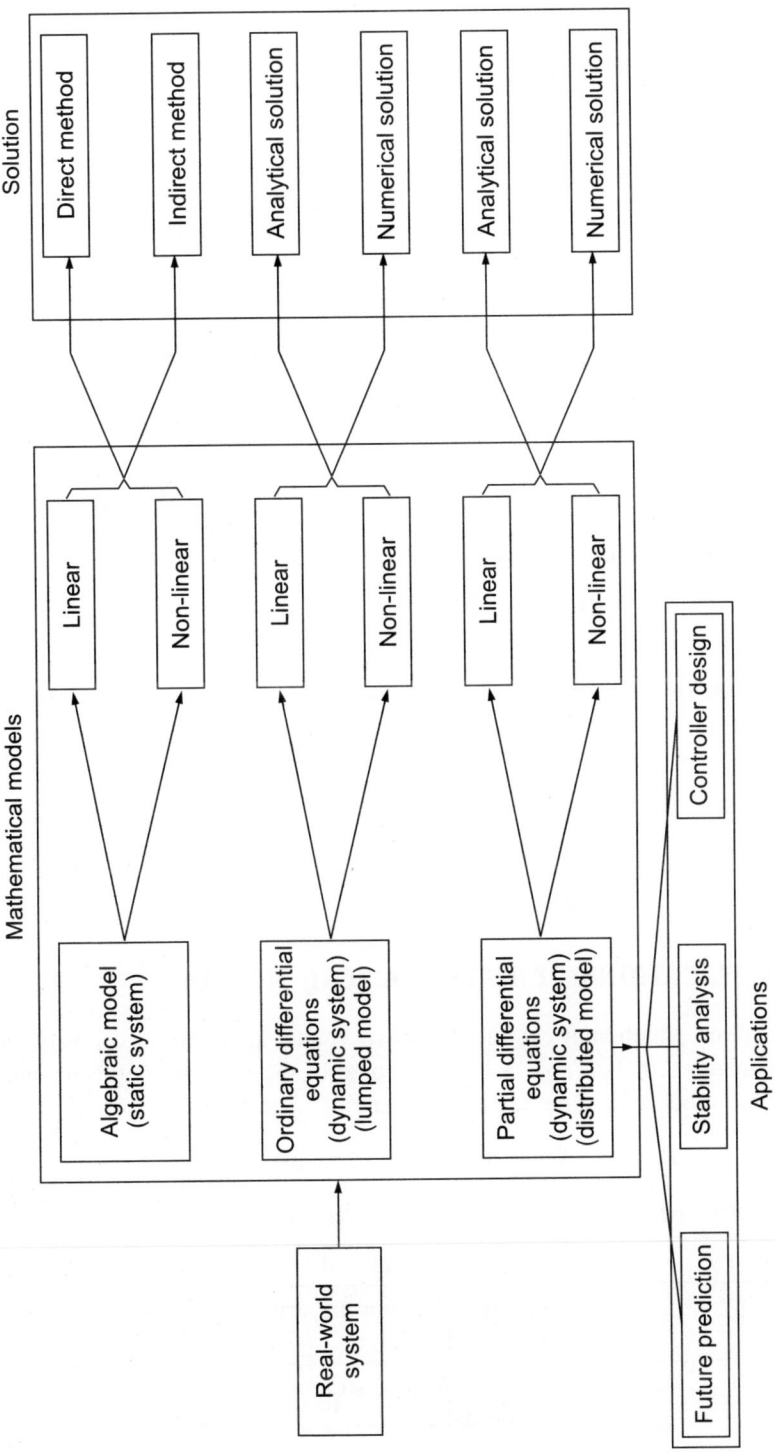

FIGURE 3.11 Various mathematical models and their applicabilities.

3.12 CONTINUOUS AND DISCRETE MODELS

A *continuous model* is a model which can produce an output at any given time, whereas, a *discrete model* is a model which can produce outputs only at specific intervals of time. This time interval is known as the *sampling period* or *sampling interval*. The process of converting a continuous model to a discrete model is referred to as the *discretisation process*. A continuous dynamic model is often represented using ordinary differential equations, whereas the discrete model is represented using difference equations. Hence, the conversion of differential equations to difference equations involves the process of discretisation. Even though representation of systems as discrete models suffers from approximation errors, it is useful to obtain the discrete model of the system for computer-based analysis and design.

In an integro-differential equation, the terms involved are the differential term $\left(\dfrac{dx}{dt}\right)$ and the integral term $\left(\displaystyle\int_0^\infty x(t)\,dt\right)$. The continuous differential term can be converted into discrete form as

$$\frac{dx}{dt} \simeq \frac{x(t) - x(t-1)}{\Delta T}, \quad t = 1, 2, \ldots$$

where ΔT is the sampling period. The continuous integral term can be converted into discrete form as

$$\int_0^\infty x(t)\,dt \simeq \sum_{n=0}^{\infty} x(n)$$

The continuous models are described using differential equations and the discrete models are described using difference equations.

3.13 APPLICATIONS OF MATHEMATICAL MODELS

In recent years, mathematical models have found applications in various fields (Neumaier, 2003). Table 3.3 shows some of the several applications of mathematical models in some areas of research.

TABLE 3.3 Applications of Mathematical Models

Field of research	Applications
Anthropology	• Classification and reconstruction of skulls
Archaeology	• Reconstruction of structures from fragments • Classification
Architecture	• Modelling of structures • Virtual reality

(Contd.)

Field of research	Applications
Artificial intelligence	• Computer vision • Pattern recognition • Decision-making
Arts	• Computer animation
Astronomy	• Detection of celestial systems • Studying the evolution of stars • Theory of the origin of the universe • Trajectory analysis of planetary systems
Biology/Biomedical engineering	• Population dynamics • Evolutionary biology • Gene analysis • DNA sequencing • Epidemiology • Genetic variability • Immune networks • Drug discovery • Medical device development • Drug testing • Synergy analysis • Bio-signal/image analysis • Neural networks • Diagnostic assistance tools • Therapy and surgery planning
Chemistry	• Analysis of chemical reaction dynamics • Molecular modelling • Analysis of chemical equilibrium • Development of chemical production systems
Criminology/forensic science	• Finger print recognition • Face recognition
Economics/finance	• Data analysis • Value estimation • Risk analysis
Electronics/electrical engineering	• Circuit design • Stability analysis • Circuit/network optimisation

(Contd.)

Field of research	Applications
Geosciences	• Generation of maps • Prediction of earthquakes • Prediction of oil deposits
Internet	• Search engines • Routing problems
Linguistics	• Automatic translation
Mechanical engineering	• Modelling of structures • Transient analysis • Frequency analysis • Vibration problems • Stability analysis • Optimisation • Study of mechanics • Thermodynamics/fluid mechanics • Crash testing
Meteorology	• Weather prediction
Music	• Synthesis and analysis of sounds
Political sciences	• Election analysis
Psychology	• Modelling of dreams/dream patterns • Modelling of human emotions
Transport science	• Scheduling traffic • Autopilot

3.14 DYNAMIC ANALYSIS

The purpose of dynamic analysis is to study and analyse the system under investigation and to perform design modifications to the system to improve its efficiency and stability. The following steps are involved in dynamic analysis:

1. Develop a simple physical model of the actual system under investigation. The developed model closely resembles the geometry and function of the actual system so that experimental analysis can be carried out.
2. Derive the mathematical model of the system using any one of the modelling approaches. The developed mathematical model closely represents the dynamics of the physical model.

3. Analyse the dynamics of the mathematical model by obtaining the solutions of the model. Further analyse the stability, catastrophes and adaptation behaviour of the system using the mathematical model.
4. Based on the results of the previous step, make design modifications to the system to improve its efficiency and performance. Design modifications include changes to the geometry and physical parameters of the system.

EXERCISES

1. For the following market model, find the value of price and quality at which market clearance is achieved.

$$\text{Supply:} \quad Q = \frac{8}{3} - \frac{2}{3}P$$

$$\text{Demand:} \quad Q = \frac{7}{2} - \frac{3}{2}P$$

2. Let X be a 6×2 design matrix, Y be a 6×1 response matrix and β be a 2×1 parameter vector. Using OLS, estimate β of the multiple regression model $Y = X\beta + \varepsilon$.

$$X = \begin{pmatrix} 1 & 0 \\ 1 & 1 \\ 1 & 2 \\ 1 & 3 \\ 1 & 4 \\ 1 & 5 \end{pmatrix}, \quad Y = \begin{pmatrix} 2 \\ 2.1 \\ 2.2 \\ 2.22 \\ 2.3 \\ 2.34 \end{pmatrix}$$

3. Fit the cubic spline for the following data given in the table below.

x	f
0	1
1	2
2	9
3	28

4. Explain how mathematical models are useful for the development of medical devices.
5. State the general assumptions commonly made during the development of dynamic models of mechanical systems.
6. Explain the mathematical models used for the prediction of climate and weather.

CHAPTER 4
Representation of a Family of Curves using Ordinary Differential Equations

Consider a family of curves with one parameter expressed as

$$f(x, y, c) = 0 \tag{4.1}$$

Equation (4.1) can be viewed as the general solution of a first-order differential equation with one parameter, which is the arbitrary constant. For various values of this parameter, one can obtain a family of curves and each of these curves is a particular solution of the given differential equation. The differential equation of the family of curves can be found by differentiating Eq. (4.1) implicitly with respect to x to obtain the following relation:

$$g\left(x, y, \frac{dy}{dx}, c\right) = 0 \tag{4.2}$$

Now, eliminating the parameter c from Eqs. (4.1) and (4.2), we can obtain

$$F\left(x, y, \frac{dy}{dx}\right) = 0 \tag{4.3}$$

which is the desired differential equation representing the given family of curves.

4.1 FAMILY OF ALL CIRCLES

Consider the family of all circles with centres at origin (0, 0), as shown in Figure 4.1.

The equation of this family is represented by
$$x^2 + y^2 = c^2 \quad (4.4)$$
Differentiating Eq. (4.4) with respect to x, we get
$$2x + 2y\frac{dy}{dx} = 0 \quad (4.5)$$
or
$$x + y\frac{dy}{dx} = 0 \quad (4.6)$$
which is the differential equation representing the family of all circles with centres at origin (0, 0).

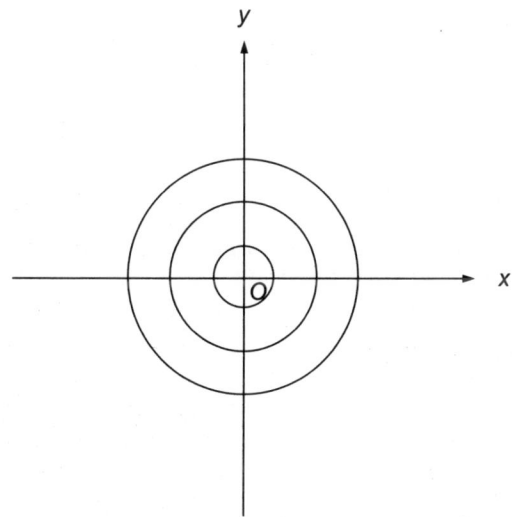

FIGURE 4.1 Family of circles.

Now, consider another example of a family of circles tangent to the x axis (Figure 4.2) given by
$$x^2 + y^2 = 2cy \quad (4.7)$$
Differentiating Eq. (4.7) with respect to x, we get
$$2x + 2y \cdot \frac{dy}{dx} = 2c\frac{dy}{dx} \quad (4.8)$$
or
$$x + (y - c)\frac{dy}{dx} = 0 \quad (4.9)$$

Now, we must eliminate c from Eq. (4.9). From the given algebraic equation, we can write

76 Mathematical Modelling of Systems and Analysis

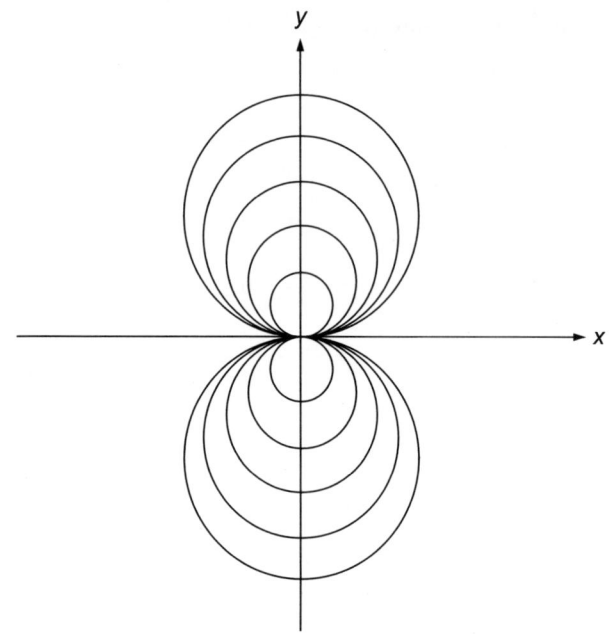

FIGURE 4.2 Family of circles tangent to the x-axis.

$$c = \frac{x^2}{2y} + \frac{y^2}{2y} \tag{4.10}$$

$$c = \frac{x^2}{2y} + \frac{y}{2} \tag{4.11}$$

Substituting Eq. (4.11) in Eq. (4.9), we get

$$x + \left(y - \frac{x^2}{2y} - \frac{y}{2}\right) \cdot \frac{dy}{dx} = 0 \tag{4.12}$$

or

$$x + \left(\frac{y}{2} - \frac{x^2}{2y}\right) \cdot \frac{dy}{dx} = 0 \tag{4.13}$$

or

$$x + \frac{1}{2}\left(\frac{y^2 - x^2}{y}\right) \cdot \frac{dy}{dx} = 0 \tag{4.14}$$

or

$$\frac{dy}{dx} = \frac{-2xy}{y^2 - x^2} \tag{4.15}$$

Equation (4.15) represents a family of circles tangent to the x-axis.

4.2 QUADRIFOLIUM CURVE

Consider a quadrifolium curve (Figure 4.3) represented by the following algebraic equation:

$$(x^2 + y^2)^3 = 4c^2 x^2 y^2 \tag{4.16}$$

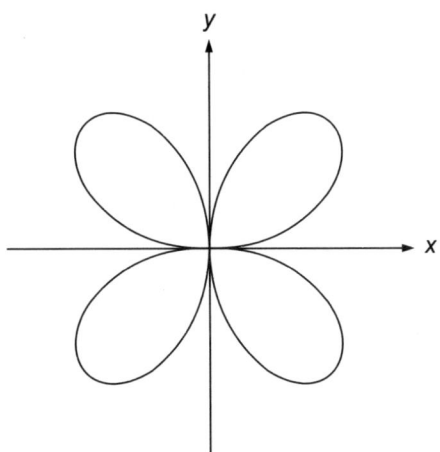

FIGURE 4.3 Quadrifolium curve.

Differentiating Eq. (4.16) with respect to x, we get

$$3(x^2 + y^2)(2x + 2yy') = 4c^2(2xy^2 + x^2 2yy') \tag{4.17}$$

or

$$4c^2 = \frac{6(x^2 + y^2)^2 (x + yy')}{2(xy^2 + x^2 yy')} \tag{4.18}$$

From Eqs. (4.17) and (4.18),

$$4c^2 = \frac{(x^2 + y^2)^3}{x^2 y^2} = \frac{3(x^2 + y^2)^2 (x + yy')}{xy^2 + x^2 yy'} \tag{4.19}$$

$$(x^2 + y^2)(xy^2 + x^2 yy') = 3x^2 y^2 (x + yy') \tag{4.20}$$

$$(x^2 + y^2)(xy^2) + (x^2 + y^2)x^2 yy' = 3x^3 y^2 + 3x^2 y^3 y' \tag{4.21}$$

$$y'(x^4 y + x^2 y^3 - 3x^2 y^3) = 3x^3 y^2 - x^3 y^2 - xy^4 \tag{4.22}$$

$$y' = \frac{dy}{dx} = \frac{xy^2 (3x^2 - x^2 - y^2)}{xy(x^3 + xy^2 - 3xy^2)} \tag{4.23}$$

$$\frac{dy}{dx} = \frac{y(2x^2 - y^2)}{x(x^2 - 2y^2)} \tag{4.24}$$

Equation (4.24) is the differential equation representing a quadrifolium curve.

4.3 TRIFOLIUM CURVE

Consider a trifolium curve (Figure 4.4) given by the following algebraic equation:

$$(x^2 + y^2)^2 = c(x^3 - 3xy^2) \tag{4.25}$$

Differentiating equation (4.25) with respect to x, we get

$$2(x^2 + y^2)(2x + 2yy') = c(3x^2 - 3y^2 - 6xyy') \tag{4.26}$$

Eliminating c from Eqs. (4.25) and (4.26), we get

$$y' = \frac{dy}{dx} = \frac{x^4 - 12y^2 x^2 + 3y^4}{6xy^3 - 10yx^3} \tag{4.27}$$

Equation (4.27) is the required equation representing a trifolium curve.

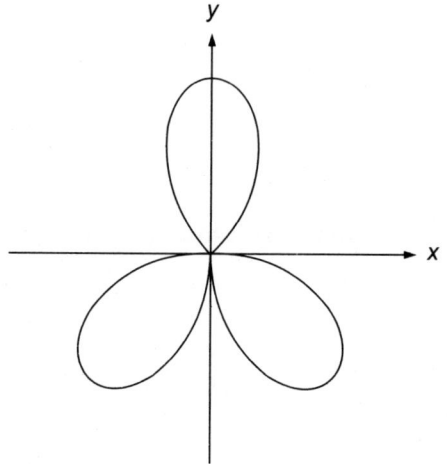

FIGURE 4.4 A trifolium curve.

Now, suppose there are two or more constants in the given algebraic equation. The differential equation can be formed by eliminating the constants in the following way:

Consider an algebraic equation.

$$y = A \cos x + B \sin x \tag{4.28}$$

where A and B are the arbitrary constants.

Differentiating Eq. (4.28) with respect to x, we get

$$\frac{dy}{dx} = -A \sin x + B \cos x \tag{4.29}$$

Differentiating Eq. (4.29) with respect to x, we get

$$\frac{d^2y}{dx^2} = -A\cos x - B\sin x = -(A\cos x + B\sin x) \qquad (4.30)$$

Now, eliminating A and B from Eqs. (4.28), (4.29) and (4.30), we get the ordinary differential equation (ODE) representing the original Eq. (4.28)

$$\frac{d^2y}{dx^2} + y = 0 \qquad (4.31)$$

Consider another example where we would like to obtain the differential equations by eliminating the constants A and B from the algebraic equation.

$$y = Ae^x + Be^{2x} \qquad (4.32)$$

Differentiating Eq. (4.32) with respect to x twice, we get

$$\frac{dy}{dx} = Ae^x + 2Be^{2x} \qquad (4.33)$$

$$\frac{d^2y}{dx^2} = Ae^x + 4Be^{2x} \qquad (4.34)$$

Now, eliminating A and B from Eqs. (4.32), (4.33) and (4.34), we get

$$\frac{d^2y}{dx^2} - 3\frac{dy}{dx} + 2y = 0 \qquad (4.35)$$

which is the differential representing the algebraic Eq. (4.32).

EXERCISES

1. Represent the family of curves $y^2 = x^2(c - x^2)$ as a differential equation.
2. The result of an experiment is given in the following table, which shows the evolution of a quantity x with respect to time t. Find the equation of the curve relating x to t and obtain the differential equation for the evolution of x with t.

Time t	x
0	0
1	1
2	4
3	9
4	16
5	25
6	36

CHAPTER 5

System Modelling using Difference Equations and Ordinary Differential Equations

Several natural and man-made systems and processes such as planetary motion, projectile motion, flow of electric current, population growth, etc. are described very well by dynamic equations such as ordinary differential equations (ODEs). The derivative of a function describes the rate at which the function changes with respect to its independent variable. Hence, the ordinary differential equations can be effectively used to model changing quantities.

5.1 TYPES OF ORDINARY DIFFERENTIAL EQUATIONS

The ODEs can be broadly classified into *linear and non-linear ODEs*. Further the linear ODE's can be classified into *homogeneous and non-homogeneous ODEs*. A homogeneous system of ordinary differential equations is of the following form:

$$\left.\begin{aligned}\frac{dx_1}{dt} &= \dot{x}_1 = a_{11}x_1 + a_{12}x_2 + \ldots + a_{1n}x_n \\ \frac{dx_2}{dt} &= \dot{x}_2 = a_{21}x_1 + a_{22}x_2 + \ldots + a_{2n}x_n \\ &\vdots \quad \vdots \\ \frac{dx_n}{dt} &= \dot{x}_n = a_{n1}x_1 + a_{n2}x_2 + \ldots + a_{nn}x_n\end{aligned}\right\}$$

For example, consider the mathematical model of love affairs presented by Strogatz in 1988. This model, known as the Romeo–Juliet model, is a simple linear homogeneous ODE model describing the love/hate relationship between the two lovers. The model is given by

$$\left.\begin{aligned}\frac{dR(t)}{dt} &= aR + bJ \\ \frac{dJ(t)}{dt} &= cR + dJ\end{aligned}\right\}$$

where, $R(t)$ is Romeo's love or hate for Juliet at time t and $J(t)$ is Juliet's love or hate for Romeo at time t. Positive values of R and J signify love and negative values signify hate. The parameters a, b, c and d may have either positive or negative sign. The choice of signs specifies their romantic styles. For example, if $a > 0$ and $b > 0$, then Romeo is an 'eager lover', who gets excited by Juliet's love for him and is further excited by his own affectionate feelings for her.

A non-homogeneous system of ODEs is of the following form:

$$\left.\begin{aligned}\frac{dx_1}{dt} &= \dot{x}_1 = a_{11}x_1 + a_{12}x_2 + \ldots + a_{1n}x_n + f_1(t) \\ \frac{dx_2}{dt} &= \dot{x}_2 = a_{21}x_1 + a_{22}x_2 + \ldots + a_{2n}x_n + f_2(t) \\ &\vdots \quad \vdots \\ \frac{dx_n}{dt} &= \dot{x}_n = a_{n1}x_1 + a_{n2}x_2 + \ldots + a_{nn}x_n + f_n(t)\end{aligned}\right\}$$

And a system of non-linear ODEs is of the following form:

$$\left.\begin{aligned}\frac{dx_1}{dt} &= \dot{x}_1 = f_1(x_1, x_2, \ldots, x_n) \\ \frac{dx_2}{dt} &= \dot{x}_2 = f_2(x_1, x_2, \ldots, x_n) \\ &\vdots \quad \vdots \\ \frac{dx_n}{dt} &= \dot{x}_n = f_n(x_1, x_2, \ldots, x_n)\end{aligned}\right\}$$

where, $f_1, f_2, \ldots, f_n$ are non-linear functions.

The above system, as mentioned in Chapter 2, is an *autonomous system of ODEs*. A *non-autonomous system of ODEs* is of the following form:

$$\left.\begin{aligned}\frac{dx_1}{dt} &= \dot{x}_1 = f_1(x_1, x_2, \ldots, x_n, t) \\ \frac{dx_2}{dt} &= \dot{x}_2 = f_2(x_1, x_2, \ldots, x_n, t) \\ &\vdots \quad \vdots \\ \frac{dx_n}{dt} &= \dot{x}_n = f_n(x_1, x_2, \ldots, x_n, t)\end{aligned}\right\}$$

The *state* of a system is defined by an independent set of position coordinates along with their derivatives.

Ordinary differential equations have been used to get insight into complex processes such as armed conflict between nations and arms race. For example, let us consider the *Richardson model* of arms race describing the phenomenon of armament levels of two countries (Richardson, 1960). Let x_1 and x_2 be the time dependent armament levels of two countries. The Richardson model is built with the following statements:

1. For each nation, its armament level will tend to increase proportionally to the armament level of the other nation.
2. For each nation, its armament level will tend to decrease proportionally with the economic burden of its own armament.
3. For each nation, its armament level will tend to increase guided by its grievances and hatred towards the other nation.

Now, the model of arms race between two nations is formulated using the following set of linear differential equations:

$$\dot{x}_1 = \sigma x_2 - \alpha x_1 + g$$
$$\dot{x}_2 = \rho x_1 - \gamma x_2 + h$$

where, σ and ρ are known as the *defence coefficients*, α and γ are known as *fatigue coefficients*, g and h are known as *grievance coefficients*.

Another model of the arms race considers both the economic as well as political aspects of arms race. This model is expressed by the set of the following differential equations:

$$\dot{x}_1 = -\beta_1 x_1 + Z_1$$
$$\dot{x}_2 = -\beta_2 x_2 + Z_2$$

where, x_1 and x_2 are the armament levels of two different countries, Z_1 and Z_2 are the expenditures on weapons at time t by each nation. The terms $\beta_1 x_1$ and $\beta_2 x_2$ ($\beta_1, \beta_2 \geq 0$) represent the resources necessary to maintain and operate the weapons x_1 and x_2, respectively. The weapons expenditure Z_i of each nation is calculated or determined by maximising the following objective functions J_i:

$$J_i = \int_0^\infty e^{-rt} U_i(Z_i, D_i(x_1, x_2)) \, dt, \quad i = 1, 2$$

where r is the discount rate, D_i is the ith country's index of defence and U_i is the instantaneous utility of ith country. Further, each nation will not have the complete freedom in allocation of its weapons' expenditures, since Z_i cannot exceed the net national product Y_i. Hence, the optimisation problem becomes a constrained optimisation problem subjected to the following constraint:

$$Z_i + C_i = Y_i, \quad i = 1, 2$$

where, C_i is the consumption of ith nation (Simaan & Cruz, 1975).

5.2 MATHEMATICAL MODELLING OF EXPONENTIAL GROWTH AND EXPONENTIAL DECAY

Consider a measured exponential growth of a variable or population with respect to time or any independent variable, as shown in Figure 5.1. A simple dynamical model of this measurement can be expressed using the linear ordinary differential equation of the following form:

$$\frac{dx}{dt} = Ax \tag{5.1}$$

The analytical solution of Eq. (5.1) is

$$x(t) = x(0) \cdot e^{At}$$

where, $x(0)$ is the initial value and A is the tuneable parameter. Hence, in order to fit the measurement data to the model, the parameter A needs to be estimated. This estimation of parameters can be performed using the method of least squares, which will be discussed later in this book.

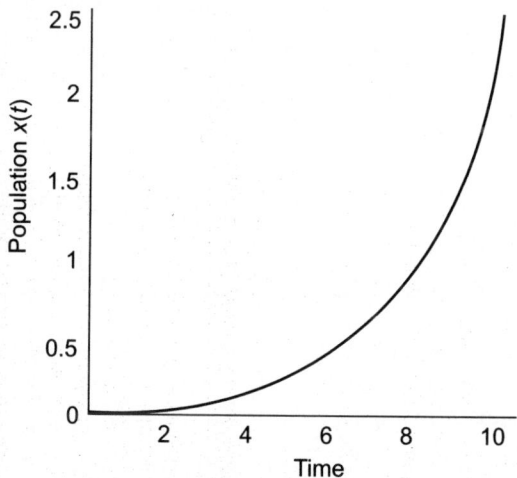

FIGURE 5.1 An exponentially increasing population.

Now, consider an exponentially decreasing population as shown in Figure 5.2. The appropriate model for such a measurement would be

$$\frac{dx}{dt} = -Ax$$

whose analytical solution is given by

$$x(t) = x(0) \cdot e^{-At}$$

For example, consider the decay of carbon-14 into nitrogen-14. This chemical equation is given by

$$C^{14} \xrightarrow{r} N^{14} + e^{-}$$

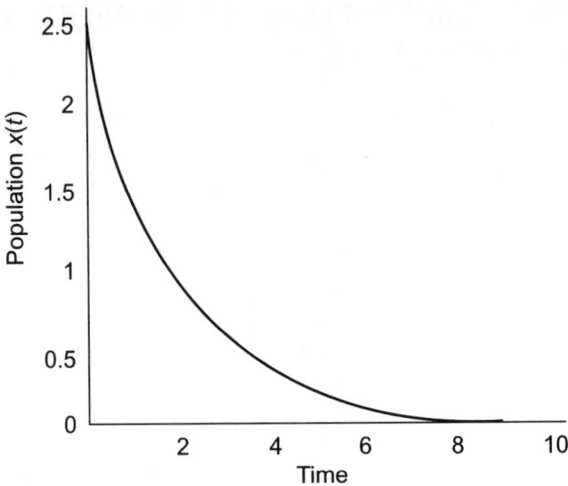

FIGURE 5.2 An exponentially decreasing population.

The rate of decay of C^{14} is proportional to the amount left and is exponentially decreasing in nature. If the rate of decay is r, then the above process can be described by the following differential equation:

$$\frac{d[C^{14}]}{dt} = -r[C^{14}]$$

where, $[C^{14}]$ represents the concentration of carbon-14 at time t.

5.3 MATHEMATICAL MODELLING OF OSCILLATORY DYNAMICS

The oscillatory dynamics of a system can be described using second order or higher order differential equations. For example, consider the mass-spring-dashpot system shown in Figure 5.3.

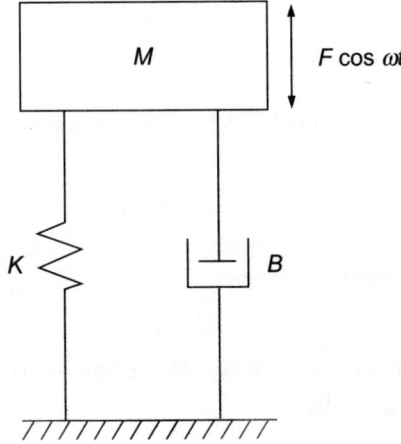

FIGURE 5.3 A mass-spring-dashpot model of a simple suspension system.

The mathematical model of this system is given by

$$M\frac{d^2x}{dt^2} + B\frac{dx}{dt} + Kx = F\cos\omega t$$

where, M, B, K, F and ω are the mass, damping coefficient, stiffness, forcing amplitude and excitation frequency, respectively. To analyse the vibration behaviour of this system, let us consider three different cases.

Case I: Free vibration ($B = 0$; $F = 0$)

For free vibration, the governing equation of the system is given by

$$M\frac{d^2x}{dt^2} + Kx = 0$$

or

$$\frac{d^2x}{dt^2} + \omega_n^2 x = 0 \tag{5.2}$$

where, $\omega_n \left(= \sqrt{\dfrac{K}{M}}\right)$ is the natural frequency of the system. The analytical solution of Eq. (5.2) can be expressed as

$$x(t) = A\cos(\omega t - \phi)$$

Case II: Free damped vibration ($F = 0$)

For free damped vibration, the equation of the system is expressed as

$$M\frac{d^2x}{dt^2} + B\frac{dx}{dt} + Kx = 0$$

or

$$\frac{d^2x}{dt^2} + 2\varsigma\omega_n\frac{dx}{dt} + \omega_n^2 x = 0 \tag{5.3}$$

where, $\varsigma \left(= \dfrac{B}{2M\omega_n}\right)$ is known as the *damping factor*. The analytical solution of Eq. (5.3) is given by

$$x(t) = e^{-\varsigma\omega_n t}\left(Ae^{\omega_n(\sqrt{\varsigma^2-1})t} + Be^{-\omega_n(\sqrt{\varsigma^2-1})t}\right)$$

If $\varsigma > 1$, then the system is overdamped; if $\varsigma = 1$, then the system is critically damped, and if $\varsigma < 1$, then the system is underdamped. The responses of free damped vibrations for different ς values are shown in Figure 5.4.

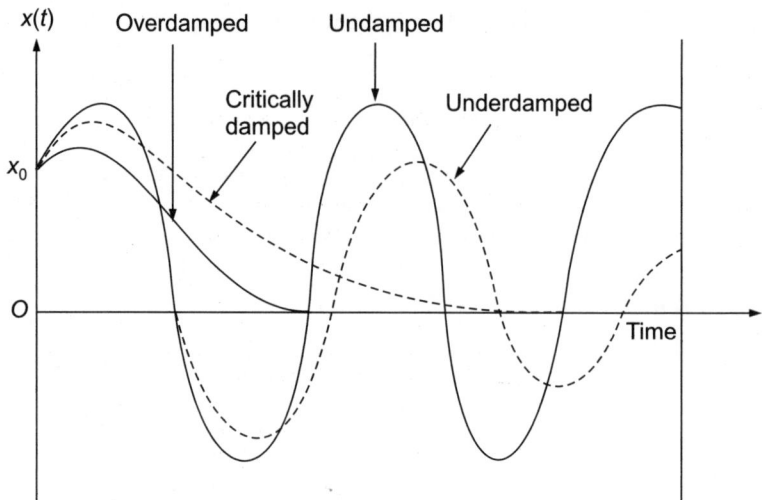

FIGURE 5.4 Responses of free damped system.

Case III: Forced vibration

For this case, the governing equation can be expressed as follows:

$$M\frac{d^2x}{dt^2} + B\frac{dx}{dt} + Kx = F\cos\omega t \text{ or } F\sin\omega t$$

or

$$\frac{d^2x}{dt^2} + 2\varsigma\omega_n\frac{dx}{dt} + \omega_n^2 x = \frac{F}{M}\cos\omega t = A\omega_n^2 \cos\omega t \quad (5.4)$$

where, A (= F/K) is known as the *displacement amplitude*.

The analytical solution of Eq. (5.4) consists of the complimentary function $x_1(t)$ and the particular integral $x_2(t)$. The complimentary function $x_1(t)$ is the same as the solutions corresponding to the free damped vibration.

$$x_1(t) = e^{-\varsigma\omega_n t}(Ae^{\omega_n(\sqrt{\varsigma^2-1})t} + Be^{-\omega_n(\sqrt{\varsigma^2-1})t})$$

and the particular integral $x_2(t)$ is given by

$$x_2(t) = X\cos(\omega t - \phi)$$

The overall solution is given by the sum of $x_1(t)$ and $x_2(t)$.

5.4 DEGREES OF FREEDOM

The degrees of freedom of a system are the number of parameters of the system that may vary independently. The models described in Section 5.3 correspond to a system with single degree of freedom. Now, consider a system, as shown in Figure 5.5, where displacement is possible in two directions.

System Modelling using Difference Equations and Ordinary Differential Equations **87**

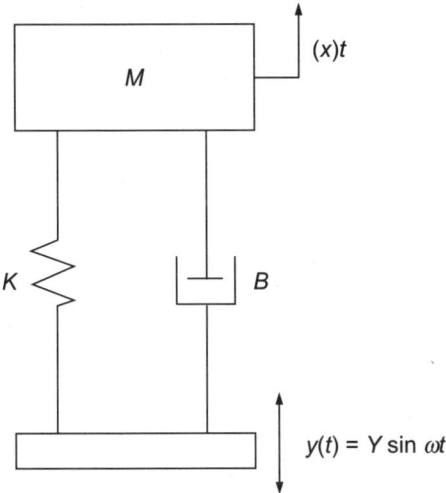

FIGURE 5.5 System with two degrees of freedom.

Now, the equation of motion is represented as

$$M\frac{d^2x}{dt^2} + B\left(\frac{dx}{dt} - \frac{dy}{dt}\right) + K(x-y) = 0$$

where, $y(t) = Y \sin \omega t$.

5.5 ORDER OF THE DIFFERENTIAL EQUATION

The order of the differential equations or the number of variables can be varied to model different real-world scenarios. The order (n) of the differential equation and its relation to the dynamics of the system is shown in Figure 5.6.

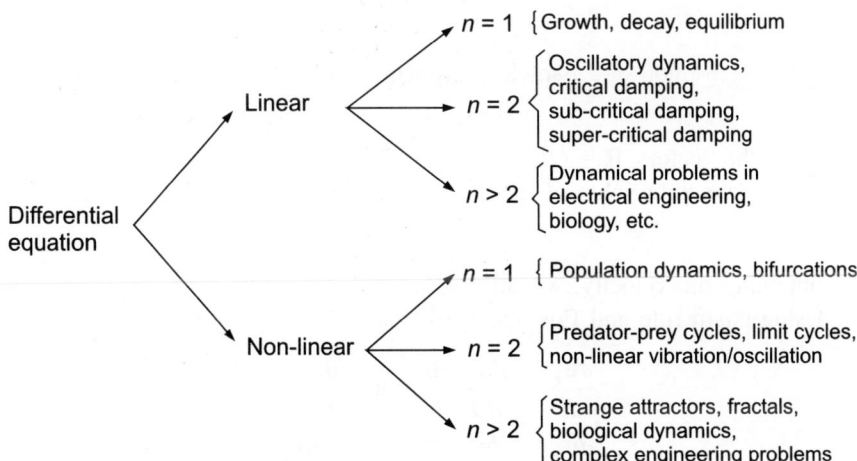

FIGURE 5.6 Relation between the order of the differential equation and the dynamics of the system.

5.6 MODELLING THE MOTION OF PLANETS AND SATELLITES

Let $P(r, \theta)$ be a particle moving on a plane curve (Figure 5.7), whose equation in polar coordinates using unit vectors $\mathbf{u}_r$ and $\mathbf{u}_\theta$ is given by

$$\left.\begin{array}{l} \mathbf{u}_r = i\cos\theta + j\sin\theta \\ \mathbf{u}_\theta = -i\sin\theta + j\cos\theta \end{array}\right\} \quad (5.5)$$

Upon differentiating Eq. (5.5) with respect to θ, we get

$$\left.\begin{array}{l} \dfrac{d\mathbf{u}_r}{d\theta} = -i\sin\theta + j\cos\theta = \mathbf{u}_\theta \\ \dfrac{d\mathbf{u}_\theta}{d\theta} = -i\cos\theta - j\sin\theta = -\mathbf{u}_r \end{array}\right\} \quad (5.6)$$

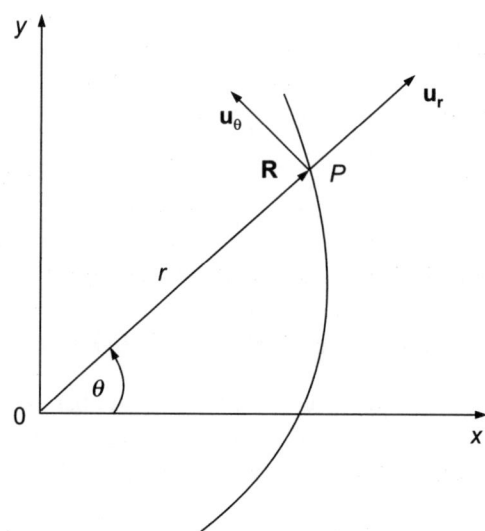

FIGURE 5.7 Motion of a particle on a plane curve.

Since the vectors $\mathbf{R} = \overrightarrow{OP}$ and $r\mathbf{u}_r$ have the same direction and the length of $\mathbf{R}$ is the absolute value of the polar coordinate r of $P(r, \theta)$, we write

$$\mathbf{R} = r\mathbf{u}_r \quad (5.7)$$

For obtaining the velocity, we differentiate w.r.t. time t.

Using chain rule and Eqs. (5.5), (5.6) and (5.7), we get

$$\left.\begin{array}{l} \dfrac{d\mathbf{u}_r}{dt} = \dfrac{d\mathbf{u}_r}{d\theta}\dfrac{d\theta}{dt} = \mathbf{u}_\theta \dfrac{d\theta}{dt} \\ \dfrac{d\mathbf{u}_\theta}{dt} = \dfrac{d\mathbf{u}_\theta}{d\theta}\dfrac{d\theta}{dt} = -\mathbf{u}_r \dfrac{d\theta}{dt} \end{array}\right\} \quad (5.8)$$

System Modelling using Difference Equations and Ordinary Differential Equations

and using Eqs. (5.7) and (5.8), we get

$$v = \frac{d\mathbf{R}}{dt} = \mathbf{u_r}\frac{dr}{dt} + r\frac{d\mathbf{u_r}}{dt} = \mathbf{u_r}\frac{dr}{dt} + \mathbf{u_\theta} r\frac{d\theta}{dt} \qquad (5.9)$$

The velocity vector must be the tangent to the curve at P and will have the magnitude

$$|v| = \sqrt{\left(\frac{dr}{dt}\right)^2 + r^2\left(\frac{d\theta}{dt}\right)^2}$$

By differentiating the velocity vector, we get the acceleration vector as

$$a = \frac{dv}{dt} = \left(\mathbf{u_r}\frac{d^2r}{dt^2} + \frac{dr}{dt}\frac{d\mathbf{u_r}}{dt}\right) + \left(\mathbf{u_\theta} r\frac{d^2\theta}{dt^2} + \mathbf{u_\theta}\frac{dr}{dt}\frac{d\theta}{dt} + \frac{d\mathbf{u_\theta}}{dt}r\frac{d\theta}{dt}\right)$$

Using Eq. (5.8),

$$a = \mathbf{u_r}\left[\frac{d^2r}{dt^2} - r\left(\frac{d\theta}{dt}\right)^2\right] + \mathbf{u_\theta}\left[r\frac{d^2\theta}{dt^2} + 2\frac{dr}{dt}\frac{d\theta}{dt}\right] \qquad (5.10)$$

In order to convert Eqs. (5.7), (5.8) and (5.10) from the xy-plane to the xyz-space, we add $\mathbf{k}z$ to the position vector, $\mathbf{k}\frac{dz}{dt}$ to the velocity vector and $\mathbf{k}\frac{d^2z}{dt^2}$ to the acceleration vector.

Hence, we have

$$\left.\begin{aligned}\mathbf{R} &= r\mathbf{u_r} + \mathbf{k}z \\ v &= \mathbf{u_r}\frac{dr}{dt} + \mathbf{u_\theta} r\frac{d\theta}{dt} + \mathbf{k}\frac{dz}{dt} \\ a &= \mathbf{u_r}\left[\frac{d^2r}{dt^2} - r\left(\frac{d\theta}{dt}\right)^2\right] + \mathbf{u_\theta}\left[r\frac{d^2\theta}{dt^2} + 2\frac{dr}{dt}\frac{d\theta}{dt}\right] + \mathbf{k}\frac{d^2z}{dt^2}\end{aligned}\right\} \qquad (5.11)$$

where, the unit vectors $\mathbf{u_r}$, $\mathbf{u_\theta}$ and $\mathbf{k}$ are mutually orthogonal. In order to analyse the motion of the planets, we consider Newton's second law and the force acting on the planet is the gravitational force towards the centre, i.e,. the sun. Let m be the mass of the planet, M be the mass of the sun and G be the gravitational constant.

$$F = ma = m\frac{d^2\mathbf{R}}{dt^2} = -\frac{GmM}{|\mathbf{R}|^2}\frac{\mathbf{R}}{|\mathbf{R}|} \qquad (5.12)$$

From Eq. (5.12), we see that the acceleration vector $\frac{d^2\mathbf{R}}{dt^2}$ is proportional to the position vector $\mathbf{R}$. Hence, their cross product is equal to 0.

$$\mathbf{R} \times \frac{d^2\mathbf{R}}{dt^2} = 0$$

Thus,
$$\mathbf{R} \times \frac{d\mathbf{R}}{dt} = C, \text{ a constant} \quad (5.13)$$

Since $\mathbf{R} \times v$ is constant, we can obtain the constant C using the cross product $\mathbf{R} \times v$ at time $t = 0$.

At $t = 0$, let $r = r_0$ and $\left(\dfrac{dr}{dt}\right)_0 = 0$. Using polar coordinates and choosing $\theta = 0$, which is the perihelion position of the planet or the position nearest to the sun at $t = 0$, we get

$$v_0 = \left(r \frac{d\theta}{dt}\right)_0 \quad (5.14)$$

Using these initial conditions and $\mathbf{u}_r \times \mathbf{u}_r = 0$, $\mathbf{u}_r \times \mathbf{u}_\theta = \mathbf{k}$,

$$C = (\mathbf{R}_0) \times (v_0) = \mathbf{k}\left(r^2 \frac{d\theta}{dt}\right)_0 = \mathbf{k}(r_0 v_0) \quad (5.15)$$

where, $\mathbf{R}_0$, v_0 and r_0 are initial conditions of $\mathbf{R}$, v and r, respectively. Substituting Eq. (5.15) in Eq. (5.13) and using Eq. (5.14), we get

$$\mathbf{k}\left(r^2 \frac{d\theta}{dt}\right) = \mathbf{k}(r_0 v_0)$$

or

$$r^2 \frac{d\theta}{dt} = r_0 v_0$$

The element of area in polar coordinates is

$$dA = \frac{1}{2} r^2 d\theta$$

and

$$\frac{dA}{dt} = \frac{1}{2} r^2 \frac{d\theta}{dt} = \frac{1}{2} r_0 v_0 = \text{Constant}$$

Hence, the line joining the planet and the sun sweeps over equal areas in equal times (Figure 5.8). This is known as *Kepler's second law*.

Kepler's first law states that the path of a planet's motion is a conic section, whose equation is

$$r = \frac{1/h}{1 + e\cos\theta}$$

where, $\dfrac{1}{h} = \dfrac{r_0^2 v_0^2}{GM}$

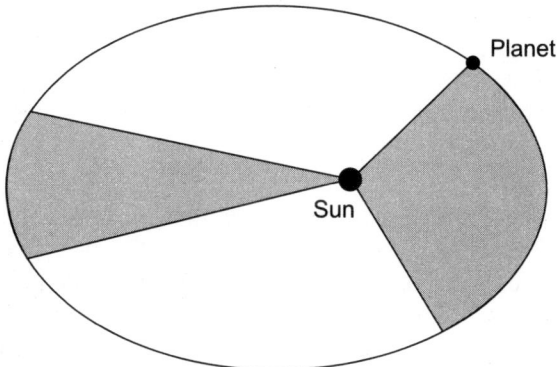

FIGURE 5.8 The line joining the planet and the sun sweeps over equal areas in equal times

and

$$e = \frac{1}{r_0 h} - 1 = \frac{r_0 v_0^2}{GM} - 1.$$

If $e = 0$, the orbit is circular. For $0 < e < 1$, the orbit is an ellipse. For $e = 1$, the orbit is a parabola. And for $e > 1$, the orbit is a hyperbola.

Kepler's third law states that when an orbit is an ellipse, the time for one period is given by

$$\frac{T^2}{a^3} = \frac{4\pi^2}{GM}$$

where, a is the length of the semi-major axis of the ellipse.

5.7 MATHEMATICAL MODELLING USING DIFFERENCE EQUATIONS

Consider the measurement of a variable x with respect to time, at a constant sampling rate. Now, it is possible to develop a recursive equation known as the *difference equation*, which represents the change in variable x. This difference equation is a dynamic model of a system and can be used to predict the changes in x in the near future. Suppose the variable $x = \{x_0, x_1, x_2, x_3, ..., x_n\}$ observed at $t = \{t_0, t_1, t_2, t_3, ..., t_n\}$, where $\Delta t = t_1 - t_0 = t_2 - t_1 = ... = t_n - t_{n-1} =$ Constant, the first differences are:

$$\Delta x_0 = x_1 - x_0$$
$$\Delta x_1 = x_2 - x_1$$
$$\Delta x_2 = x_3 - x_2$$
$$\vdots$$
$$\Delta x_{n-1} = x_n - x_{n-1}$$

In general, the nth first difference is

$$\Delta x_n = x_{n+1} - x_n$$

Now, the difference equation model of the system is given by

$$\Delta x_n = x_{n+1} - x_n = f(x_n)$$

where, $f(\bullet)$ can be a linear or a non-linear function. Modelling of systems using difference equations now involves finding the function $f(\bullet)$, which would describe the change in x in an appropriate manner. The modelling methodology for difference equations can be explained through the following examples.

EXAMPLE 5.1 Using the observations of a variable x given in the following Table, develop a suitable difference equation model.

n	x_n
0	2.1
1	4.41
2	9.261
3	19.4481
4	40.841
5	85.7661

Solution: Calculating the first differences,

$$\Delta x_0 = x_1 - x_0 = 4.41 - 2.1 = 2.31$$
$$\Delta x_1 = x_2 - x_1 = 9.261 - 4.41 = 4.851$$
$$\Delta x_2 = x_3 - x_2 = 19.4481 - 9.261 = 10.1871$$
$$\vdots$$

By plotting Δx_n as a function of x_n, we can see that the relationship between the change in x_n or Δx_n with x_n is linear, as shown in the following figure. Hence, the difference equation is given by

$$\Delta x_n = x_{n+1} - x_n = k x_n$$

where, $k = \dfrac{\Delta x_n}{x_n} = 1.1$

and the model is given by

$$x_{n+1} = 1.1 x_n + x_n$$

with initial condition $x_0 = 2.1$

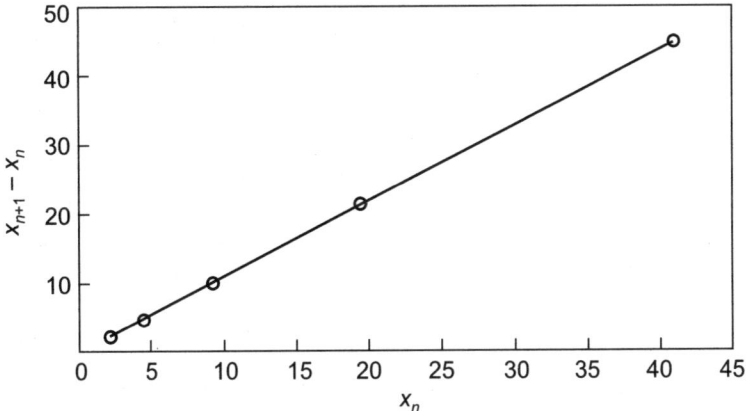

EXAMPLE 5.2 Develop a suitable difference equation model for the observations of a variable x given in the following table.

n	x_n
0	1
1	3
2	21
3	903
4	1631721

Solution: Calculating the first differences, we get

$$\Delta x_0 = x_1 - x_0 = 2$$
$$\Delta x_1 = x_2 - x_1 = 18$$
$$\Delta x_2 = x_3 - x_2 = 882$$
$$\vdots$$

By plotting and observing the change Δx_n as a function of x_n, we see that the relationship is quadratic in nature. Hence, the difference equation is

$$\Delta x_n = x_{n+1} - x_n = f(x_n) = k x_n^2$$

and,
$$k = \frac{\Delta x_n}{x_n^2} = 2$$

The required model is

$$x_{n+1} = 2x_n^2 + x_n \quad \text{with } x_0 = 1$$

For more complex measurement sets, the relationship between Δx_n and x_n can be mapped using higher order polynomial models using computer-aided curve fitting tools.

A difference equation model is a dynamic model and can be viewed as an infinite set of algebraic equations.

5.8 INTEGRO-DIFFERENTIAL EQUATIONS

In general, an integro-differential equation is of the form

$$\frac{dx(t)}{dt} = f(t, x(t)) + \int_0^t g(t, s, x(s))ds, \quad x(0) = x_0$$

These equations find applications in modelling of complex population dynamics, electrical engineering, mechanical engineering, etc. A few examples of integro-differential equation models in population biology are given in Chapter 8.

5.9 DELAY DIFFERENTIAL EQUATIONS

Several processes in the fields of biology, engineering, economics, etc. involve differential equations with time delays. For example, consider the growth of a population of a particular species modelled using ordinary differential equations. In such a case, it is realistic to introduce a time delay into the model, since the members of the population must reach a threshold age before giving birth to a new individual. Such effects are efficiently modelled using delay differential equations. In general, a delay differential equation is of the following form:

$$\frac{dx(t)}{dt} = f(t, x(t), x(t - T))$$

For example, consider the exponentially growing population with a time delay

$$\frac{dx}{dt} = ax(t - T), \quad t > T$$

In this model, we see that the rate of change of the population at time t depends on the population at time $(t - T)$. A few examples of delay differential equations in modelling the population dynamics are presented in Chapter 8.

5.10 DIFFERENTIAL-DIFFERENCE EQUATIONS

If $u_n(t)$, $n = 1, 2, ...$, is a function of t, having a derivative

$$\frac{d}{dt} u_n(t) = u'_n(t),$$

an equation involving $u'_n(t)$ and $u_n(t)$, $u_{n+1}(t)$, $u_{n+2}(t)$, ... is called a *differential-difference equation* or a *recursive differential equation*. For example, $x'_n(t) = ax_n(t) + bx_{n-1}(t) + cx_{n+1}(t)$ is a differential-difference equation.

5.11 FRACTIONAL ORDER DIFFERENTIAL EQUATIONS

Fractional calculus generalises the integration and differentiation to non-integer order fundamental operator $_aD_t^\alpha$, where a and t are the limits of the operation and $\alpha \in R$, and is defined as

$$_aD_t^\alpha = \begin{cases} \dfrac{d^\alpha}{dt^\alpha} & : \alpha > 0 \\ 1 & : \alpha = 0 \\ \int_a^t (d\tau)^{-\alpha} & : \alpha < 0 \end{cases}$$

Some of the commonly used definitions for the fractional differential are as follows:

1. Grunwald–Letnikov definition

$$_aD_t^\alpha f(t) = \lim_{h \to 0} h^{-\alpha} \sum_{j=0}^{\left[\frac{t-a}{h}\right]} (-1)^j \binom{\alpha}{j} f(t - jh)$$

2. Riemann–Liouville definition

$$_aD_t^\alpha f(t) = \frac{1}{\Gamma(n-\alpha)} \frac{d^n}{dt^n} \int_a^t \frac{f(\tau)}{(t-\tau)^{\alpha-n+1}} d\tau, \quad (n-1 < \alpha < n)$$

3. Caputo definition

$$_aD_t^\alpha f(t) = \frac{1}{\Gamma(\alpha-n)} \int_a^t \frac{f^{(n)}(\tau)}{(t-\tau)^{\alpha-n+1}} d\tau, \quad (n-1 < \alpha < n)$$

4. Consider a linear fractional order differential equation of the form

$$\frac{d^\alpha x(t)}{dt^\alpha} = D_t^a x(t) = bx(t), \quad x(0) = x_0$$

Its solution is given by

$$x(t) = x_0 E_{\alpha,1}(bt^\alpha)$$

where, $E_{\alpha,1}$ is the Mittag–Leffler function.

$$E_{\alpha,1}(z) = \sum_{k=0}^{\infty} \frac{z^k}{\Gamma(\alpha k + 1)}$$

where, Γ denotes the *gamma* function. For modelling examples, see Chapter 13.

5.12 STOCHASTIC DIFFERENTIAL EQUATION

It can be imagined that probabilistic or non-deterministic models are more appropriate and realistic than the models which are deterministic in nature. Several real-world phenomena can be viewed as a random phenomena in addition to dynamic determinism. Hence, it is often necessary to analyse random variables, which evolve as functions of time or space. *Stochastic processes* are those processes which involve families of random variables that are functions of time or space. For example, consider a variable measured as a function of time t (Figure 5.9). However, in reality, most of the experimentally measured values include a random effect affecting the system (Figure 5.10).

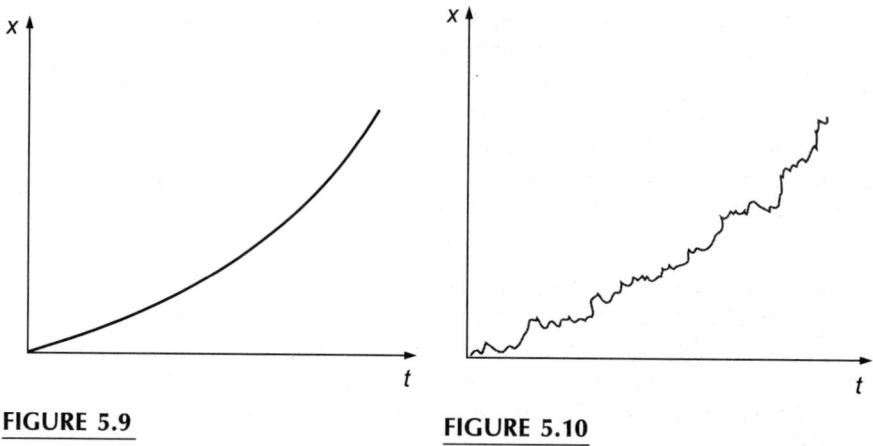

FIGURE 5.9   FIGURE 5.10

Stationary process: If for $t_1, t_2, t_3, \ldots, t_n$, the joint distribution of the vector random variables $(X(t_1), X(t_2), X(t_3), \ldots, X(t_n))$ and $(X(t_1+h), X(t_2+h), X(t_3+h), \ldots, X(t_n+h))$ are the same for all $h > 0$, then the stochastic process $\{X(t), t \in T\}$ is said to be stationary of order n.

Gaussian process: If the distribution of $(X(t_1), X(t_2), X(t_3), \ldots, X(t_n))$ for all $t_1, t_2, t_3, \ldots, t_n$ is multivariate normal, then $X(t)$ is said to be a Gaussian process.

Markov process: If $\{X(t), t \in T\}$ is a stochastic process such that given the value $X(s)$, the values of $X(t), t > s$ do not depend on the values of $X(u), u < s$, then the process is called a Markov process.

Wiener process: A standard Wiener process $W = \{W(t), t \geq 0\}$ on $[0, T]$, is a process with stationary independent increments such that for any $0 \leq t_1 < t_2 \leq T$, the increment $W(t_2) - W(t_1)$ is a Gaussian random variable with mean zero and variance equal to $t_2 - t_1$, i.e., $E[W(t_2) - W(t_1)] = 0$; $\text{Var}[W(t_2) - W(t_1)] = t_2 - t_1$.

Consider an ordinary differential equation expressed as

$$\dot{x}(t) = ax(t)$$

One way of including the effects of randomness is by expressing the above system in the following way:

$$\dot{x}(t) = ax(t) + bx(t)r(t)$$

where, $r(t)$ can be a random process such as a white noise.

Now, consider a general form of a stochastic differential equation:

$$dX = f(X, t)dt + g(X, t)dW$$

where, W is a standard Wiener process and X is a random process. $f(\bullet)$ and $g(\bullet)$ are real-valued functions. The solution of this system is given by

$$X(t) = X(0) + \int_0^t f(X(t'), t')\, dt' + \int_0^t g(X(t'), t')\, dW(t')$$

As an example from financial modelling, consider the model for the evolution of a bank account B. The bank account evolves according to the following stochastic differential equation:

$$\frac{dB(t)}{dt} = r(t)B(t)$$

Suppose $B(0) = 1$. Then, investing unit value of currency will be at a value of $B(t)$ at time t. The stochastic variable $r(t)$ describes the instantaneous yield on the bank account and is called *short rate*. Stochastic differential equations find major applications in the fields of finance and economics.

EXERCISES

1. What are the advantages and applications of modelling a system using ordinary differential equations?
2. Formulate an ODE model for love triangles. Use the Romeo–Juliet model as the starting model.
3. For a mass-spring-damper system with $M = 3$, $B = 2$ and $K = 3$, calculate the damping factor, sketch the response and comment on the type of the response of the free damped system.
4. With the help of a suitable example, explain the degrees of freedom of a system.

CHAPTER 6

Mathematical Modelling using Balance Laws

The principles of conservation of mass and energy can be utilised to obtain the ODE models of various systems. The law of conservation of mass states that for a closed system, the mass of the system remains constant over time. This law implies that mass can neither be created nor be destroyed; however it can be rearranged in space. For example, in a chemical reaction, the total mass of the reactants or starting materials must be equal to the mass of the products. Similarly, the law of conservation of energy states that energy cannot be created or destroyed, however, it changes from one form to another.

However, in special relativity, a principle called *mass–energy equivalence* states that the mass of a system is a measure of its energy content. The equivalence of energy E and mass M is dependent on the speed of light c and is given by the equation $E = Mc^2$.

6.1 MATHEMATICAL MODELLING USING MATERIAL BALANCE

Consider an example of a cylindrical tank system, as shown in Figure 6.1. Let F_{in} be the inflow to the tank, F_{out} be the outflow from the tank and h be the level of the liquid inside the tank. The mathematical model of this system can be obtained using material balance equation, i.e.,

$$\begin{pmatrix} \text{Rate of change} \\ \text{of volume of} \\ \text{liquid in the tank} \end{pmatrix} = \begin{pmatrix} \text{Inflow to} \\ \text{the tank} \end{pmatrix} - \begin{pmatrix} \text{Outflow} \\ \text{from} \\ \text{the tank} \end{pmatrix}$$

Mathematically, this can be represented as

$$\frac{dV}{dt} = F_{in} - F_{out}$$

The volume of the liquid in the cylindrical tank can be expressed as

$$V = \pi r^2 h$$

where, r is the radius of the tank and h is the liquid level (height) in the tank.

Since, the cross-sectional area A ($= \pi r^2$) is always constant, the governing equation can be expressed as

$$A \frac{dh}{dt} = F_{in} - F_{out}$$

Suppose, there is a valve regulating the outflow and the coefficient of the valve is represented as b, then the outflow is related to the liquid level as

$$F_{out} = b\sqrt{h}$$

Hence,

$$A \frac{dh}{dt} = F_{in} - b\sqrt{h} \qquad (6.1)$$

Equation (6.1) is a non-linear equation having a square root type non-linearity.

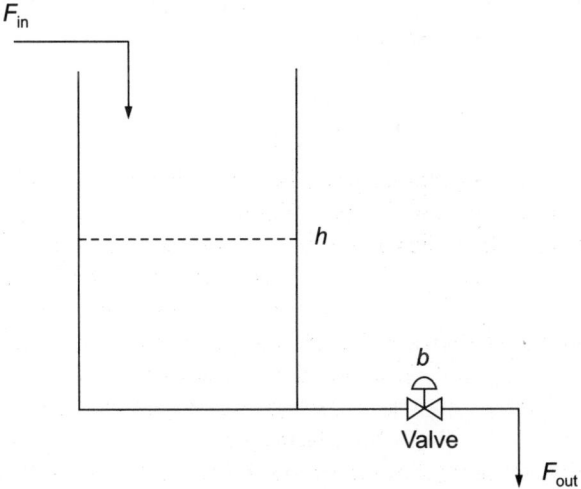

FIGURE 6.1 A single tank system.

6.2 MODELLING USING ENERGY BALANCE

6.2.1 MODELLING OF THE PROCESS OF DRILLING HOLES USING LASER

The cutting, welding and drilling of metals is of major importance in several areas of engineering and technology. For this process, high-powered lasers are very useful. The idea behind this is to focus a lot of power on a small area of the surface of a metal to produce surface heating and evaporation, leading to the subsequent formation of a hole (Andrews and McLone, 1976).

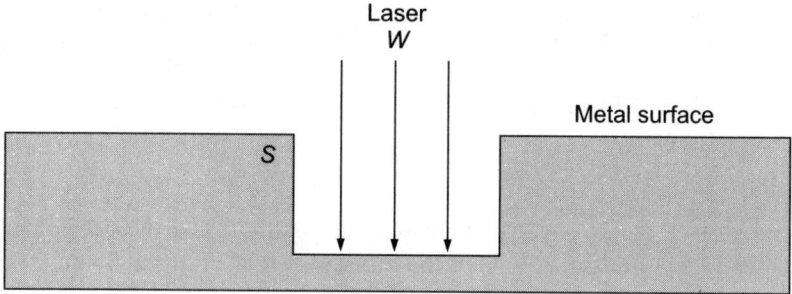

FIGURE 6.2 Drilling a hole using laser.

Consider that the energy applied through the laser at the surface is used to evaporate the material. Assume that the laser power W is distributed uniformly over some area of the surface A and is applied perpendicular to the surface (Figure 6.2). In a time interval δt, the amount of energy dissipated is $W\delta t$ and the depth of the hole increases by δS. Hence, the volume of material evaporated is $A\delta S$, and by the law of conservation of energy,

$$h\rho A \delta S = W \delta t$$

where, h is the heat required to vaporise unit mass of material and ρ is the density of the material. The model of the drilling process can now be written as

$$\frac{dS}{dt} = \frac{W/A}{h\rho}$$

where W/A is the power density.

Integrating the differential equation and setting $S = 0$ at $t = 0$, the depth of the hole at any instant t is

$$S(t) = \frac{1}{h\rho A} \int_0^t W dt$$

$$= \frac{E(t)}{h\rho A}$$

where, $E(t)$ is the total energy dissipated by the laser source in the time interval $(0, t)$.

6.2.2 Hamilton's Method

Hamilton's method is efficient for obtaining the ODE model of conservative systems, where the equations of motion are obtained using the law of conservation of energy (Hamilton, 1834).

To illustrate this method, let us try to obtain the ODE model of the free fall of an object of mass M, falling under the influence of gravity only. Let $x(t)$ be the height of the object from the ground. For this system, we have

Kinetic energy $= \dfrac{1}{2} M v^2$

Potential energy $= Mgx$

The total energy of this system is

$$H = \text{Kinetic energy} + \text{Potential energy}$$

$$= \dfrac{1}{2} M v^2 + Mgx$$

where, H denotes the total energy, which corresponds to the Hamiltonian of the system. For a conservative system, we have

$$\dfrac{dH}{dt} = 0$$

$$\dfrac{dH}{dt} = \dfrac{1}{2} M 2v \cdot \dfrac{dv}{dt} + Mg \dfrac{dx}{dt} = 0 \quad (6.2)$$

Since $v = \dfrac{dx}{dt}$ and $\dfrac{dv}{dt} = \dfrac{d^2 x}{dt^2}$, from Eq. (6.2), we have

$$\dfrac{1}{2} M \cdot 2 \cdot \dfrac{dx}{dt} \cdot \dfrac{d^2 x}{dt^2} + M \cdot g \cdot \dfrac{dx}{dt} = 0$$

$$M \cdot \dfrac{dx}{dt} \left(\dfrac{d^2 x}{dt^2} + g \right) = 0$$

Hence, $$\dfrac{d^2 x}{dt^2} = -g$$

which is the ODE model of the system.

Now, consider another example where we would like to obtain the ODE model of the system shown in Figure 6.3.

Let x be the displacement, M be the mass and K be the stiffness of the spring. We know that the model of the system is $M \dfrac{d^2 x}{dt^2} + Kx = 0$, when no external force is applied. Let us check whether we obtain the same equation using the Hamilton's method.

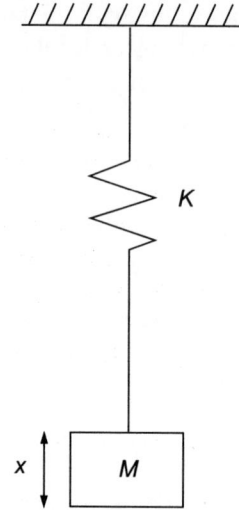

FIGURE 6.3 A mass-spring system.

For this system, we have

$$\text{Kinetic energy} = \frac{1}{2}Mv^2$$

and

$$\text{Potential energy} = \frac{1}{2}Kx^2$$

The total energy of the system is

$$H = \frac{1}{2}Mv^2 + \frac{1}{2}Kx^2$$

$$\frac{dH}{dt} = \frac{1}{2}M.2v.\frac{dv}{dt} + \frac{1}{2}K.2x.\frac{dx}{dt}$$

For the conservative system, $\frac{dH}{dt} = 0$.

Hence,

$$\frac{1}{2}M.2v.\frac{dv}{dt} + \frac{1}{2}K.2x.\frac{dx}{dt} = 0$$

Since $v = \frac{dx}{dt}$ and $\frac{dv}{dt} = \frac{d^2x}{dt^2}$, we get

$$M.\frac{dx}{dt}.\frac{d^2x}{dt^2} + Kx.\frac{dx}{dt} = 0$$

or
$$\frac{dx}{dt}\left(M \cdot \frac{d^2x}{dt^2} + Kx\right) = 0$$

$$\Rightarrow M\frac{d^2x}{dt^2} + kx = 0$$

Consider another example where we model the pendulum motion under the influence of gravity alone, as shown in Figure 6.4. Let M be the mass of the bob and l be the length of the rod and consider the air resistance to be negligible.

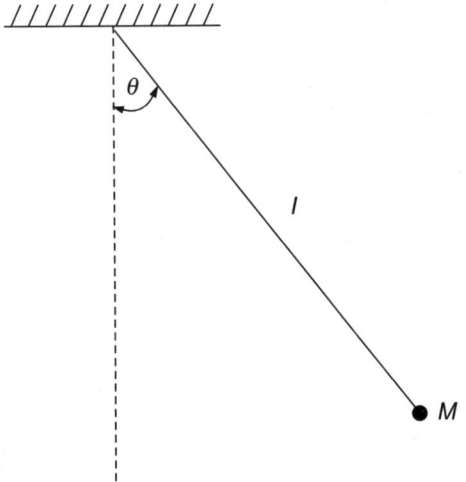

FIGURE 6.4 Pendulum motion under the influence of gravity.

The kinetic energy of the system is given by

$$\text{KE} = \frac{1}{2}Mv^2 = \frac{1}{2}Ml^2\left(\frac{d\theta}{dt}\right)^2$$

where the distance is the arc length $s = l\theta$.

Hence,
$$v = \frac{ds}{dt} = l \cdot \frac{d\theta}{dt}$$

The potential energy is given by
$$\text{PE} = MgL$$

where, L is the vertical distance between the pendulum's position and its height when hanging straight down.

$$L = l - l\cos\theta$$

Hence, the total energy of the system is given by

$$H = KE + PE$$

$$= \frac{1}{2}Ml^2\left(\frac{d\theta}{dt}\right)^2 + Mgl(1 - \cos\theta)$$

Setting $\dfrac{dH}{dt} = 0$, we get

$$Ml^2 \cdot \frac{d\theta}{dt} \cdot \frac{d^2\theta}{dt^2} + Mgl\sin\theta \frac{d\theta}{dt} = 0$$

$$\Rightarrow \frac{d^2\theta}{dt^2} = \frac{-g}{l}\sin\theta$$

6.2.3 THE TAUTOCHRONE

We shall now discuss a much more complicated modelling problem involving the law of conservation of energy. Consider a wire bent like a smooth curve and let a bead of mass M starts from point P and slide down to the origin without any friction (Figure 6.5).

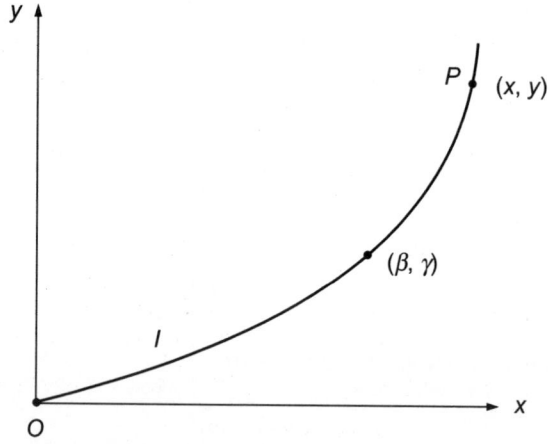

FIGURE 6.5

Suppose (x, y) is the starting point and (β, γ) is any intermediate point and the shape of the wire is given by a function $y = y(x)$. Then, the total time of descent will be a definite function $T(y)$ of the initial height y. Now, if we specify the function $T(y)$ in advance, we can then find the curve or the shape of the wire that gives $T(y)$. The required curve is called the *tautochrone*. Let l be the length of the arc of the curve from the origin O. By the law of conservation of energy, we can equate the gain in kinetic energy to the loss of potential energy, and hence, we get

$$\frac{1}{2}M\left(\frac{dl}{dt}\right)^2 = Mg(y-\gamma)$$

Thus,
$$\frac{dl}{dt} = -\sqrt{2g}\sqrt{y-\gamma}$$

or
$$dl = -\sqrt{2g}\sqrt{y-\gamma}\, dt$$

By separating the variables and integrating from $\gamma = y$ to $\gamma = 0$, we get

$$T(y)\sqrt{2g} = \int_{\gamma=0}^{\gamma=y} \frac{dl}{\sqrt{y-\gamma}} \qquad (6.3)$$

where, T is the time of descent.

Now, let $l = H(\gamma)$, then $dl = H'(\gamma)\, d\gamma$ and the function $H(\gamma)$ depends upon the curve. Eq. (6.3) can be written as

$$T(y)\sqrt{2g} = \int_0^y (y-\gamma)^{-1/2} H'(\gamma)\, d\gamma \qquad (6.4)$$

Eq. (6.4) can be written as

$$T(y)\sqrt{2g} = \int_0^y (y-\gamma)^{-1/2} r(\gamma)\, d\gamma \qquad (6.5)$$

where $r(y) = H'(y)$. From Eq. (6.5), we get

$$T(y) = \frac{1}{\sqrt{2g}} \int_0^y \frac{r(\gamma)\, d\gamma}{(y-\gamma)^{1/2}}$$ which is called as the Abel's integral equation.

Hence,

$$T(y)\sqrt{2g} = y^{-1/2} * H'(y) \qquad (6.6)$$

where, $*$ denotes the convolution product.

Applying Laplace transform for Eq. (6.6), we get

$$L[T(y)] = \frac{1}{\sqrt{2g}} L[y^{-1/2}] L[H'(y)]$$

But $L[H'(y)] = sL(H(y)) - H(0) = sh(s)$, since $H(0) = 0$.

Since $L[y^{-1/2}] = \sqrt{\dfrac{\pi}{s}}$, we get

$$L[H'(y)] = \sqrt{2g} \frac{L[T(y)]}{\sqrt{\pi/s}}$$

$$= \sqrt{\frac{2g}{\pi}} s^{1/2} L[T(y)]$$

If $T(y)$ is a constant T_0, then

$$L[H'(y)] = \frac{\sqrt{2g}}{\pi} s^{1/2} L[T_0]$$

$$= \sqrt{\frac{2g}{\pi}} s^{1/2} \frac{T_0}{s}, \quad \text{since } L(1) = 1/s$$

$$= a^{1/2} \sqrt{\frac{\pi}{s}} \qquad (6.7)$$

where, $a = \dfrac{2gT_0^2}{\pi^2}$.

By taking inverse Laplace transform of Eq. (6.7), we get

$$r(y) = \sqrt{\frac{a}{y}} \qquad (6.8)$$

where, $r(y) = H'(y)$

Now, $H(y) = \displaystyle\int_0^y \sqrt{1 + \left(\dfrac{dx}{dy}\right)^2}\, dy$ is the length of the curve and its derivative

$$r(y) = H'(y) = \sqrt{1 + \left(\frac{dx}{dy}\right)^2} \qquad (6.9)$$

Substituting Eq. (6.9) in Eq. (6.8), we get

$$1 + \left(\frac{dx}{dy}\right)^2 = \frac{a}{y}$$

which is the differential equation of the curve. By separating the variables and integrating, we get

$$dx = \sqrt{\frac{a-y}{y}}\, dy \qquad (6.10)$$

Substituting, $y = a \sin^2 \dfrac{\theta}{2}$ in Eq. (6.10), we get

$$dx = \sqrt{\frac{a-y}{y}}\,dy = \sqrt{\frac{a-a\sin^2(\theta/2)}{a\sin^2(\theta/2)}}\;2a\sin(\theta/2)\cos(\theta/2)\frac{1}{2}d\theta$$

$$= \frac{\cos(\theta/2)}{\sin(\theta/2)}\,a\sin(\theta/2)\cos(\theta/2)\,d\theta = a\cos^2(\theta/2)\,d\theta$$

$$= \frac{a(1+\cos\theta)}{2}\,d\theta$$

Integrating, we get, $x = \dfrac{a}{2}(\theta + \sin\theta)$, since $x = 0$ when $y = 0$.

But, $y = a\sin^2\dfrac{\theta}{2} = \dfrac{a}{2}(1-\cos\theta)$ and hence we get the parametric equation of the *tautochrone* as

$$x = \frac{a}{2}(\theta + \sin\theta) \text{ and } y = \frac{a}{2}(1-\cos\theta)$$

which represents one arch of the cycloid generated by a point P on a circle of radius $a/2$. Since $a = \dfrac{2gT_0^2}{\pi^2}$, the diameter of the generating circle is determined by the time T_0 of descent.

EXERCISE

1. For the system shown in the following figure, derive the mathematical model for the rate of change of volume of liquid in each tank.

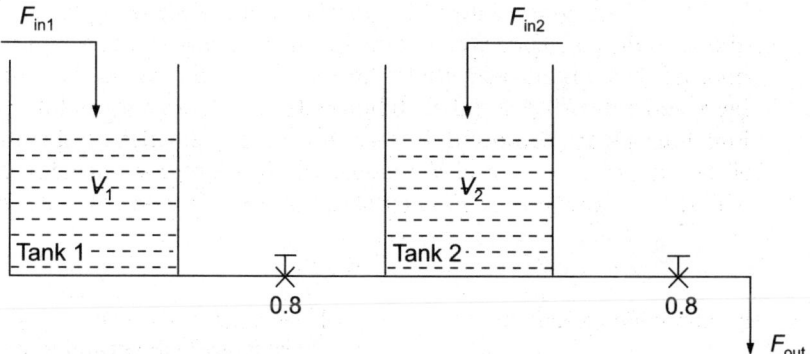

Assume the tanks to be cylindrical and F_{in1} is the inflow rate of liquid to tank 1 and F_{in2} is the inflow rate of liquid to tank 2. F_{out} is the outflow from tank 2. The liquid can flow from tank 1 to tank 2 or from tank 2 to tank 1 through the interconnecting pipe whose valve coefficient is 0.8. V_1 and V_2 are volumes of tank 1 and tank 2, respectively.

CHAPTER 7
Mathematical Modelling of Chemical Reactions

All chemical reactions take place at a definite rate depending on the conditions such as concentrations of the reacting substances, temperature, pressure, catalyst and radiation. Such conditions are known as *experimental conditions*. The subject of reaction kinetics involves the study of the rates of chemical reactions and the influence of these conditions on the rates.

It is well established that the rate of a chemical reaction is proportional to the concentrations of the reacting substances since they are continuously consumed in the course of the reactions. Since their concentrations decrease steadily, the rate of reaction must vary with time, as shown in Figure 7.1, and the process becomes slower and slower as the reactants are used up.

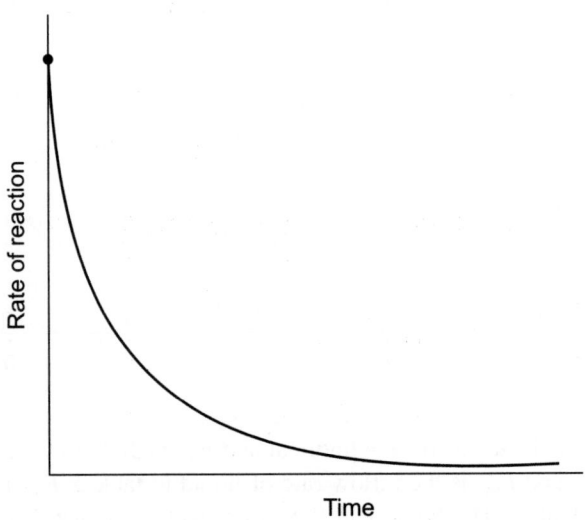

FIGURE 7.1

Chemical reactions can be classified based on the order of reaction (i.e., by the number of atoms or molecules whose concentrations determine the rate of reaction). The rate of a chemical reaction is proportional to the product of the n_1th power of concentration C_1 of reactant 1, the n_2th power of concentration C_2 of reactant 2, and so on.

$$\text{Rate} = kC_1^{n_1} C_2^{n_2} C_3^{n_3} \ldots$$

$$= k \prod_{i=1}^{m} C_i^{n_i}$$

The reaction is said to be of overall order n, where $n = n_1 + n_2 + n_3 + \ldots$ and there are no reactions which are higher than the third order.

The rate equations of chemical reactions can be conveniently represented using the first order differential equations (Glasstone and Lewis, 1983).

7.1 FIRST ORDER REACTION

In a first order reaction, the rate is directly proportional to the concentration of one reactant.

$$\frac{-dC}{dt} = KC \tag{7.1}$$

where, C is the concentration of the reactant at time t and K is the proportionality constant known as the velocity constant or rate constant or the specific reaction rate. The negative sign indicates that C decreases while t increases.

By the method of separation of variables, Eq. (7.1) is rearranged into

$$\frac{-dC}{C} = Kdt \tag{7.2}$$

Upon integration, we get

$$-\int_{C_0}^{C} \frac{dC}{C} = K \int_{0}^{t} dt$$

or

$$\ln \frac{C_0}{C} = Kt$$

and

$$K = \frac{1}{t} \ln \frac{C_0}{C} \tag{7.3}$$

Now, consider that a moles of reactant are initially present in unit volume of the reacting system and x is the number of moles in unit volume that have reacted after time t, leaving $a - x$ unreacted. Then, Eq. (7.3) becomes

$$K = \frac{1}{t} \ln \frac{a}{a-x} = \frac{2.303}{t} \log \frac{a}{a-x}$$

A typical example of a first order reaction is the gaseous reaction, namely, the decomposition of nitrogen pentoxide, given by

$$N_2O_5 \xrightarrow{K} N_2O_4 + \frac{1}{2}O_2$$

7.2 SECOND ORDER REACTION

The rate of a second order reaction depends on two concentration terms:

Case I

Consider the general case of a reaction involving two molecules of the same reactants yielding products.

$$2A \xrightarrow{K} \text{Products}$$

or

$$A + A \xrightarrow{K} \text{Products}$$

The rate of this reaction can be given as

$$\frac{-dC_A}{dt} = KC_A^2 \qquad (7.4)$$

where C_A is the concentration of reactant A at time t. The equation can be generalised to a reaction of nth order involving a single reactant as

$$\frac{-dC}{dt} = KC^n \qquad n \geq 2 \qquad (7.5)$$

Separating the variables,

$$\frac{-dC}{C^n} = K dt \qquad (7.6)$$

Integrating from the limits of t from 0 to t and the limits of C from C_0 to C, we get

$$K = \frac{1}{(n-1)t}\left(\frac{1}{C^{n-1}} - \frac{1}{C_0^{n-1}}\right) \qquad (7.7)$$

For a second order reaction, $n = 2$ and Eq. (7.7) becomes

$$K = \frac{1}{t}\left(\frac{1}{C} - \frac{1}{C_0}\right) \qquad (7.8)$$

Now, let $a = C_0$ be the initial condition of the reactant and $a - x$ be the concentration at time t. Thus, Eq. (7.8) becomes

$$K = \frac{1}{t}\left(\frac{1}{a-x} - \frac{1}{a}\right) = \frac{1}{t} \cdot \frac{x}{a(a-x)}$$

Case II

Consider a second order reaction involving one molecule of two different reactants A and B.

$$A + B \xrightarrow{K} \text{Products}$$

Now, the reaction rate is given by

$$\frac{-dC_A}{dt} = \frac{-dC_B}{dt} = KC_A C_B \quad (7.9)$$

Let a and b represent the initial molar quantities of A and B, respectively, in a given volume and x be the quantity of each that has reacted after time t. Then the concentrations

$$C_A = a - x$$

and

$$C_B = b - x$$

and Eq. (7.9) becomes

$$\frac{dx}{dt} = K(a-x)(b-x) \quad (7.10)$$

or

$$\int_0^x \frac{dx}{(a-x)(b-x)} = K \int_0^t dt$$

Using partial fractions, $\dfrac{1}{(a-x)(b-x)} = \dfrac{1}{(a-b)}\left[\dfrac{1}{b-x} - \dfrac{1}{a-x}\right]$

Hence, $\displaystyle\int_0^x \frac{1}{(a-b)}\left[\frac{1}{b-x} - \frac{1}{a-x}\right] dx = K \int_0^t dt$

Upon integration,

$$\frac{1}{(a-b)} \ln \frac{b(a-x)}{a(b-x)} = Kt$$

or

$$K = \frac{1}{t(a-b)} \ln \frac{b(a-x)}{a(b-x)}$$

$$= \frac{2.303}{t(a-b)} \log \frac{b(a-x)}{a(b-x)}$$

A typical example of second order reaction involves the reaction of hydrogen and ethylene to yield ethane.

$$H_2 + C_2H_4 \xrightarrow{K} C_2H_6$$

7.3 THIRD ORDER REACTION

A third order reaction involves three molecules whose concentrations change in a course of the reaction. For example,

$$A + B + C \xrightarrow{K} \text{Products}$$

$$2A + B \xrightarrow{K} \text{Products}$$

$$3A \xrightarrow{K} \text{Products}$$

Now, consider the general case, where

$$A + B + C \xrightarrow{K} \text{Products}$$

The rate of reaction at any instant is given by

$$\frac{-dC_A}{dt} = \frac{-dC_B}{dt} = \frac{-dC_C}{dt} = KC_A C_B C_C \quad (7.11)$$

where C_A, C_B and C_C are concentrations of reactants A, B and C, respectively. If $A = B = C$, then Eq. (7.11) becomes

$$\frac{-dC}{dt} = KC^3$$

On integrating, we get

$$K = \frac{1}{2t}\left(\frac{1}{C^2} - \frac{1}{C_0^2}\right)$$

Replacing C_0 by a and C by $a - x$, we get

$$K = \frac{1}{2t}\left(\frac{1}{(a-x)^2} - \frac{1}{a^2}\right)$$

Now, let a, b, and c denote the initial conditions of A, B and C, respectively, in a given volume and x be the quantity of each that has reacted after time t. Then, the concentrations

$$C_A = a - x$$
$$C_B = b - x$$
$$C_C = c - x$$

and Eq. (7.11) becomes:

$$\frac{dx}{dt} = K(a-x)(b-x)(c-x) \quad (7.12)$$

For a reaction of the type:

$$2A + B \longrightarrow \text{Products}$$

The rate equation is

$$\frac{dx}{dt} = K(a-2x)^2(b-x)$$

On integrating, we get

$$K = \frac{1}{t(2b-a)^2}\left[\frac{(2b-a)2x}{a(a-2x)} + \ln\frac{b(a-2x)}{a(b-x)}\right]$$

7.4 OPPOSING REACTION

Consider the case where the process is reversible and forward and reverse reactions occur simultaneously. Such reaction is represented by

$$A \underset{K'}{\overset{K}{\rightleftharpoons}} B$$

where, K and K' are specific rates of forward and reverse reactions, respectively, and the ratio K/K' is equal to the equilibrium constant for the given system.

Let a be the concentration of A in the beginning of the reaction, and initially, there is no B present. After a period of time t, the concentrations of A and B will be $a - x$ and x respectively. Hence, the net rate of the reaction is given as

$$\frac{dx}{dt} = K(a-x) - K'x \tag{7.13}$$

where, $K(a - x)$ is the rate of forward reaction and $K'x$ is the rate of reverse reaction. At equilibrium,

$$K(a-x_E) = K'x_E \tag{7.14}$$

where, x_E is the amount of B formed or the amount of A decomposed, at equilibrium.

$$K' = \frac{K(a-x_E)}{x_E} \tag{7.15}$$

Substituting Eq. (7.15) in Eq. (7.13), we get

$$\frac{dx}{dt} = K(a-x) - \frac{Kx}{x_E}(a-x_E)$$

$$= \frac{Ka}{x_E}(x_E - x)$$

On integration with conditions $x(0) = 0$ and $x(t) = x$, we obtain

$$\frac{Ka}{x_E} = \frac{1}{t}\ln\frac{x_E}{x_E - x}$$

114 Mathematical Modelling of Systems and Analysis

or
$$K = \frac{x_E}{at} \ln \frac{x_E}{x_E - x}$$

Now, consider a complex reversible reaction in gaseous state, involving the dissociation of hydrogen iodide.

$$2HI \underset{K'}{\overset{K}{\rightleftharpoons}} H_2 + I_2$$

Let the initial concentration of hydrogen iodide be a and x be the extent of decomposition after time t. Then, the concentration of hydrogen iodide at t will be $a - x$ and the concentrations of both hydrogen and iodine will be $\frac{1}{2}x$.

Now, the net rate of the reaction is given by

$$\frac{dx}{dt} = K(a-x)^2 - K'\left(\frac{1}{2}x\right)^2$$

7.5 CONSECUTIVE REACTION

Consider a reaction involving three or more molecules taking place consecutively in a series of successive stages.

$$A \xrightarrow{K_1} B \xrightarrow{K_2} C$$

where K_1 and K_2 are the rate constants.

Let a be the initial concentration of the reactant A and C_A, C_B and C_C are the concentrations of A, B and C, respectively, with $a = C_A + C_B + C_C$, i.e., the total concentration is constant.

The decomposition of A is a first order process, and hence, the rate equation for A can be written as

$$\frac{-dC_A}{dt} = K_1 C_A \tag{7.16}$$

Since $C_A(0) = a$ and $C_A(t) = C_A$, on integration, we get

$$C_A(t) = ae^{-K_1 t} \tag{7.17}$$

Now, the rate at which the concentration of B increases is equal to the difference between the rate of its formation from A, i.e., $K_1 C_A$ and the rate of its decomposition into C, which is $K_2 C_B$. Hence, we can write:

$$\frac{-dC_B}{dt} = K_1 C_A - K_2 C_B \tag{7.18}$$

Integrating Eq. (7.18) and using Eq. (7.17), we get

$$C_B(t) = a \frac{K_1}{K_2 - K_1} (e^{-K_1 t} - e^{-K_2 t})$$

EXERCISES

1. Write the differential equation of the following chemical reaction and calculate the rate constant:

$$N_2O_5(g) \xrightarrow{K} N_2O_4(g) + \frac{1}{2}O_2(g)$$

2. What is the order of the following reaction:

$$CH_3COOCH_3 + H_2O \longrightarrow CH_3COOH + CH_3OH$$

Further, compute the rate constant of this reaction.

3. Derive the ODE model of the hydrogenation of ethylene to ethane.

$$C_2H_4 + H_2 \longrightarrow C_2H_6$$

CHAPTER 8

Mathematical Modelling of Population Dynamics

The modelling of population dynamics using ODEs provides a useful method for analysis of population growth over a period of time. Further, such models are useful for predicting the population after a period of time. For example, consider a pond in which the variation in the population of fish as a function of time needs to modelled. In such a case, the logistic model provides a useful representation of such population dynamics.

8.1 STEADY POPULATION

If a population increases steadily with respect to time, it is clear that the first time derivative of the population is constant. In such a case, the mathematical model is given by

$$\frac{dx}{dt} = C, \ x(0) = x_0$$

where, x is the total members of a population at time t, and x_0 is the initial population at time $t = 0$. The analytical solution of the above model is

$$x(t) = Ct + x_0$$

8.2 EXPONENTIALLY GROWING OR DECREASING POPULATION

As discussed earlier, an exponentially growing population is by

$$\frac{dx}{dt} = ax, \ x(0) = x_0$$

An exponentially decreasing population is given by

$$\frac{dx}{dt} = -ax, \ x(0) = x_0$$

where, a is the growth or decay rate.

In reality, several cases follow an exponentially increasing or decreasing population dynamics.

8.3 MALTHUSIAN MODEL

This model was developed by a British economist, Thomas R. Malthus. In this model, it is assumed that both the birth rate of a population and the death rate of a population are proportional to the current size of the population. This model is represented as follows:

$$\frac{dx}{dt} = bx - dx, \ x(0) = x_0$$

where, b is the birth rate and d is the death rate.

8.4 THE LOGISTIC MODEL

In natural systems, most populations have a limiting size beyond which their environment can no longer sustain them. In the population models described in the Sections 8.1, 8.2 and 8.3, such a limitation is not considered. This problem is overcome by the logistic model developed by the Belgian Mathematician, Pierre Verhulst in 1838. The logistic model incorporates a factor known as the carrying capacity K, which is the greatest population an environment can sustain. The logistic model is given by

$$\frac{dx}{dt} = (bx - dx)\left(1 - \frac{x}{K}\right); \ x(0) = x_0$$

The analytical solution of this model is obtained as

$$x(t) = \frac{x_0 K}{(K - x_0) e^{-(b-d)t} + x_0}$$

8.5 LOTKA–VOLTERRA MODEL

Vito Volterra was an Italian Mathematician who developed the predator-prey model, which describes the scenario, where one species of animal survives by eating another species that tries to avoid being eaten.

Let us imagine a closed geographical location inhabited by foxes and rabbits. The foxes eat rabbits and rabbits eat vegetation. Assuming that the rabbits always have enough supply of food and when the rabbits are abundant, then the foxes flourish and their population grows. When the number of foxes becomes large and eat too many rabbits, then they enter a period of famine and their population begins to decrease. As the fox population decreases, the rabbits become relatively safe and the rabbit population starts to increase again. Due to this, the fox population again starts to increase. Hence, there is a repeated cycle of increase and decrease in the population of the two species (Figure 8.1).

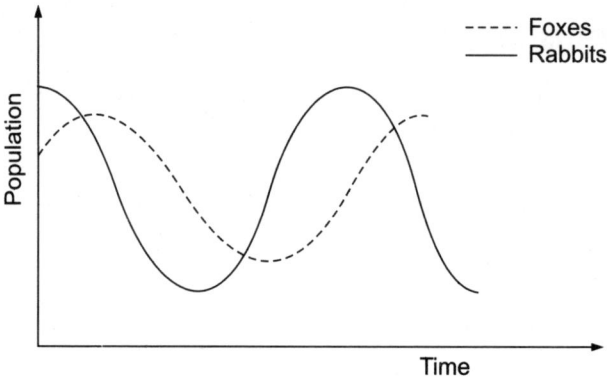

FIGURE 8.1 A typical variation of fox and rabbit population in a closed geographical location.

Suppose x is the number of rabbits at time t, and there is unlimited supply of vegetation for the rabbits and the number of foxes is zero. Then,

$$\frac{dx}{dt} = ax \quad a > 0$$

Now, if we assume that the number of encounters per unit time between rabbits and foxes is jointly proportional to x and y and that a certain proportion of these encounters results in a rabbit being eaten. Then,

$$\frac{dx}{dt} = ax - bxy \quad a, b > 0$$

and

$$\frac{dy}{dt} = -cy + dxy \quad c, d > 0$$

Hence, we have the following non-linear model describing the interaction between these two species as

$$\left.\begin{array}{l}\dfrac{dx}{dt} = x(a - by) \\ \dfrac{dy}{dt} = -y(c - dx)\end{array}\right\}$$

where, a, b, c and d are model parameters. This model is known as the Lotka–Volterra's predator-prey model.

8.6 COMPETITION MODELS

In some cases, there may be two species such as rabbits and deer that may compete for the same resources for their growth. Such cases are not the same as the predator-prey interactions discussed in Section 8.5. In case where two species use the same resources for their growth and development, the interaction is modelled using a class of models known as the *competition models*. Such a model is represented by

$$\left.\begin{aligned}\frac{dx}{dt} &= ax\left(1 - \frac{x+y}{K}\right) \\ \frac{dy}{dt} &= by\left(1 - \frac{x+y}{K}\right)\end{aligned}\right\}$$

where, x and y are populations of the competing species, K is the carrying capacity of the environment, a and b are the growth rates of species 1 and species 2, respectively. If the total population $(x + y)$ exceeds the carrying capacity, both populations will begin to die.

8.7 EPIDEMIC MODEL: THE SIR MODEL FOR THE SPREAD OF DISEASE

Epidemiology is the study of patterns, causes and effects of diseases in fixed populations. The SIR model is an epidemiological model, which computes the theoretical number of people infected with a contagious disease in a closed population of individuals over time. The term SIR comes from the fact that the model involves three different populations, namely, the number of susceptible people $S(t)$, the number of infected people $I(t)$ and the number of people who have recovered from the disease $R(t)$.

For obtaining the SIR model, the dependent variables are as follows:

$S [= S(t)]$ is the number of susceptible individuals.

$I [= I(t)]$ is the number of infected individuals.

$R [= R(t)]$ is the number of recovered individuals.

If N is the total fixed population in a closed area, then the fraction of the total population in each of the above three categories can be represented as

$s(t) \left[= \dfrac{S(t)}{N}\right]$ is the susceptible fraction of the population.

$i(t) \left[= \dfrac{I(t)}{N}\right]$ is the infected fraction of the population.

$r(t) \left[= \dfrac{R(t)}{N}\right]$ is the recovered fraction of the population.

Now, ignoring new births and deaths, and assuming that the total population N is always constant,

$$S(t) + I(t) + R(t) = N$$

and

$$s(t) + i(t) + r(t) = 1$$

Consider that on average, each infected individual generates $b.s(t)$ new infected individuals and the rate of change of the susceptible population depends on the number of individuals already susceptible, the number of individuals already infected and the amount of contact between susceptible and infected individuals. Hence, the susceptible equation can be expressed as

$$\frac{ds}{dt} = -b.s(t).i(t)$$

and if a fixed fraction k of the infected group will recover during any time of the day, the recovered equation can be expressed as

$$\frac{dr}{dt} = k.i(t)$$

Now, since, $s(t) + i(t) + r(t) = 1$, then $\dfrac{d(s+i+r)}{dt} = 0$.

Hence,

$$\frac{ds}{dt} + \frac{di}{dt} + \frac{dr}{dt} = 0$$

or

$$\frac{di}{dt} = -\frac{ds}{dt} - \frac{dr}{dt}$$

Hence, the infected equation can be expressed as

$$\frac{di}{dt} = b.s(t)i(t) - k.i(t)$$

Finally, the SIR model is expressed as the set of simultaneous differential equations as follows:

$$\frac{ds}{dt} = -bs(t)i(t)$$

$$\frac{di}{dt} = bs(t)i(t) - k.i(t)$$

$$\frac{dr}{dt} = ki(t)$$

and $s(0)$, $i(0)$ and $r(0)$ are the initial populations of susceptible, infected and recovered people, respectively. For different diseases, the parameters of the system (b and k) can be tuned and the epidemic models for different diseases can be developed. Figure 8.2 shows a typical evolution of the susceptible population, infected population and recovered population as a function of time.

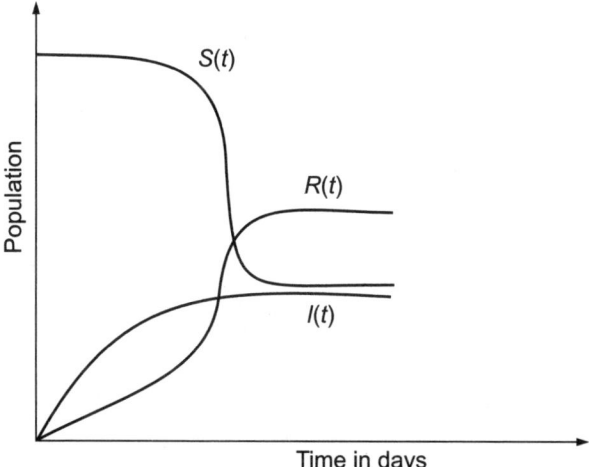

FIGURE 8.2   Typical evolutions of susceptible, infected and recovered populations.

8.8 MODELLING OF HIV-IMMUNE SYSTEM INTERACTION

Now, let us consider a challenging modelling example. The dynamics of the interaction between the human immunodeficiency virus is rather complex and is represented using a predator-prey type model.

The dynamics of the HIV infection in the human body is primarily due to the interaction of the HIV with the T-lymphocytes of the human immune system. There are two types of T-lymphocytes, namely, the T-helper cells (CD4+ cells) and T-cytotoxic cells (CD8+ cells). When these cells are activated, the T-cytotoxic cells can eliminate the host cells that are infected with virus and other parasites (Chamberlain, 2010). The HIV infection results in the selective depletion of CD4 T-cells. Since, CD4 cells are highly important in immune regulation, the decrease in CD4 cell count leads to a degrading effect on the functionality of the immune system (Perelson et al., 1993).

Exposure to the human immunodeficiency virus results in the initial local replication of the virus within target cells of the mucosal tissue (Douek et al., 2003). When HIV infects the human body, it targets the CD4 T-cells. A protein (GP120) present on the surface of the virus has a high affinity for the CD4 protein on the surface of the T-cell. Hence, binding takes place, and the contents of the HIV are injected into the host T-cell. After the DNA of the virus has been duplicated by the host cell, it is reassembled and new virus particles emerge from the surface of the host cell (Levy, 1994; Kirschner, 1996). This process can either take place in a slow and gradual manner, sparing the host cell; or in a fast and rapid fashion, bursting and killing the host CD4 cell. Further, the infected CD4 cells, simulate the CD8 cells which kill the infected CD4 T-helper cells (Wang et al., 2004). This process is presented as a block diagram in Figure 8.3.

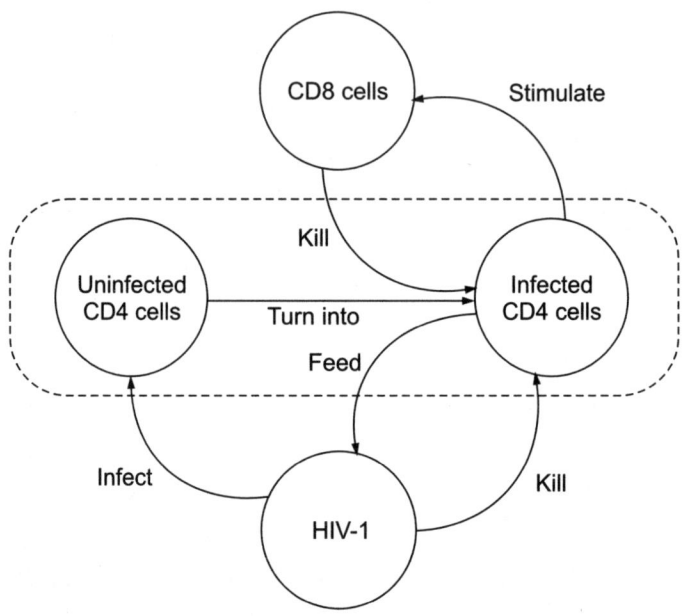

FIGURE 8.3 Interaction between HIV-1 and the human immune system.

The response of the concentrations of the CD4 lymphocyte population, the CD8 lymphocyte population and the HIV-1 viral load can be characterised by the following simultaneous first order non-linear differential equations (Menezes Campello de Souza, 1999; Ho and Ling, 2010):

$$\frac{dx(t)}{dt} = a(x_0 - x(t)) - bx(t)z(t)$$

$$\frac{dy(t)}{dt} = c(y_0 - y(t)) + dy(t)z(t)$$

$$\frac{dz(t)}{dt} = z(t)(ex(t) - fy(t))$$

where, $x(t)$, $y(t)$ and $z(t)$ are the concentrations of the CD4 lymphocyte population, the CD8 lymphocyte population and the HIV-1 viral load, respectively. x_0 and y_0 are the normal unperturbed concentrations of the CD4 and CD8 lymphocyte population. a, b, c, d, e and f are the system parameters (Ge et al., 2005). The description of each parameter is provided in Table 8.1.

TABLE 8.1 Description of HIV Model Parameters

Parameter	Description
a	Death rate of CD4 cells
b	Rate of infection of CD4 cells by virus
c	Death rate of CD8 cells
d	Rate of increase in CD8 cells in response to increased viral load
e	Rate of increase in viral load
f	Rate of decrease in viral load

Further, the dynamics of the model can be described as follows:

1. The growth rate of CD4 diminishes when the HIV-1 population grows.
2. The growth rate of HIV-1 increases with the increase in the population of HIV-1 and CD4.
3. The growth rate of HIV-1 decreases with the decrease in the population of HIV-1 and CD8.

8.9 DELAY DIFFERENTIAL EQUATIONS IN MODELLING OF POPULATION DYNAMICS

Logistic equation with time delay

A time delay can be introduced in the logistic equation (discussed in Section 8.4) in the following way:

$$\frac{dx}{dt} = (bx(t) - dx(t))\left[1 - \frac{x(t-T)}{K}\right]$$

Lotka–Volterra predator-prey model as a delay differential equation

$$\left.\begin{array}{l}\dot{x} = ax(t)\left[\dfrac{K - x(t)}{K}\right] - bx(t)y(t) \\ \dot{y} = -cy(t) + dx(t-T)y(t-T)\end{array}\right\}$$

8.10 POPULATION MODELLING USING INTEGRO-DIFFERENTIAL EQUATIONS

Mathematical model of a single population

$$\frac{dx(t)}{dt} = x(t)\left[a - bx(t) - \int_0^t k(t-s)x(s)\,ds\right]$$

where, x(t) is the population of the species at time t, a and b are positive constants and k(t) is the hereditary influence. If k(t) = 0, the model reduces to the logistic equation.

Predator-prey model

Consider a population of predator x_1 and prey x_2 evolving as a function of time. The growth and interaction of the two species is described by

$$\left. \begin{aligned} \frac{dx_1(t)}{dt} &= \left[c - a_1 x_2(t) - \int_{-\tau}^{0} b_1(x_2(t+s))ds \right] x_1(t) \\ \frac{dx_2(t)}{dt} &= \left[-c + a_2 x_1(t) - \int_{-\tau}^{0} b_2(x_1(t+s))ds \right] x_2(t) \end{aligned} \right\}$$

where, c, a_i and b_i are parameters of the system, which represents the interaction of the two species.

8.11 POPULATION MODELLING USING DIFFERENCE EQUATIONS

The heroin addiction problem

As an example for discrete modelling, consider a closed geographical location in which the problem of heroin addiction is growing. The total population is divided into three classes, namely, the susceptibles (S), the addicted (A) and the population undergoing rehabilitation (R). The susceptibles are the people who are not yet addicted to heroin, but have potential to become addicted. The addicted population is the population already addicted and R represents the population that was previously addicted but has now joined the rehabilitation programme. Let P_{SA} be the probability with which the susceptible population enters into addicted population, and P_{AR} be the probability with which the addicted population may join rehabilitation. The people under rehabilitation may recover from the addiction and enter the susceptible class with a probability P_{RS}. Few addicted persons may fight the addiction and enter the susceptible class with a probability P_{AS} and few people undergoing rehabilitation may be unable to fight addiction and may re-enter the addicted class with a probability P_{RA} (Figure 8.4).

Suppose if S_0, A_0 and R_0 are the initial populations in the susceptible, addicted and rehabilitation classes, respectively, then we can arrive at the following model, using Figure 8.4.

$$\left. \begin{aligned} S_1 &= P_{AS} A_0 + P_{RS} R_0 \\ A_1 &= P_{SA} S_0 + P_{RA} R_0 \\ R_1 &= P_{AR} A_0 \end{aligned} \right\}$$

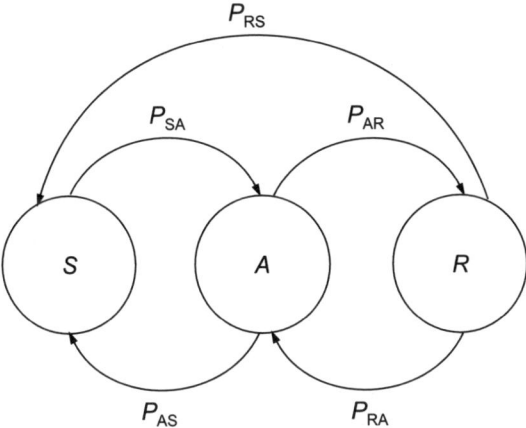

FIGURE 8.4 Transitions in the heroin addicted, susceptible and rehabilitated population categories.

More generally, we can write these equations as follows:

$$\left.\begin{aligned} S_{t+1} &= P_{AS}A_t + P_{RS}R_t \\ A_{t+1} &= P_{SA}S_t + P_{RA}R_t \\ R_{t+1} &= P_{AR}A_t \end{aligned}\right\}$$

Now, it may be convenient to think that a fraction of population from each class of population may die out. In such a case, let us consider that P_{Sd}, P_{Ad} and P_{Rd} are the probabilities with which the susceptibles, the addicted and the people under rehabilitation may die out (Figure 8.5). We now add loss terms to each of the equations in the previously derived model to obtain the following set of equations:

$$\left.\begin{aligned} S_{t+1} &= P_{AS}A_t + P_{RS}R_t - P_{Sd}S_t \\ A_{t+1} &= P_{SA}S_t + P_{RA}R_t - P_{Ad}A_t \\ R_{t+1} &= P_{AR}A_t - P_{Rd}R_t \end{aligned}\right\}$$

The three-level laser

Consider a three-level lasing process. In this process, the atoms are raised or pumped from the ground level 1 to higher energy level 3. After an atom has been raised to the higher energy level 3, it decays rapidly to level 2. After this, a population inversion exists between levels 2 and 1. Hence, the number of atoms in level 2 will be greater than the number of atoms in level 1. Now, the atoms from the level 2 go down to level 1 by releasing photons known as the laser light, as shown in Figure 8.6.

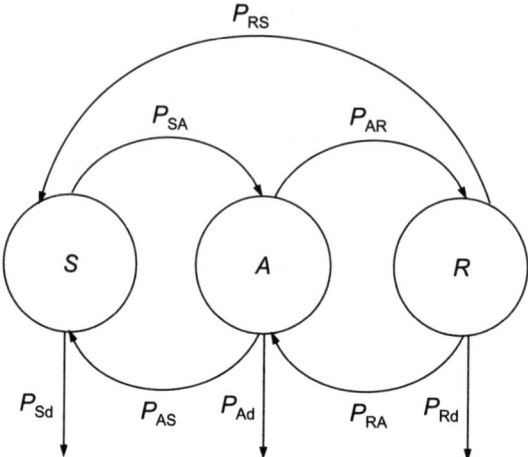

FIGURE 8.5 Transitions in the heroin addicted, susceptible and rehabilitated population categories including deaths.

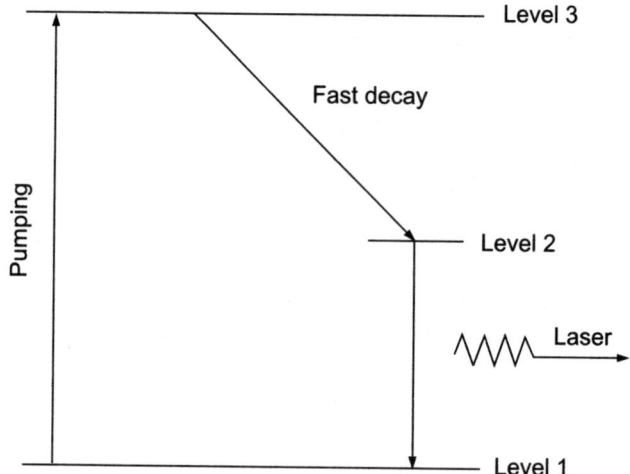

FIGURE 8.6 The three-level laser process.

We now try to develop the discrete dynamic model of this process. Let $N1$, $N2$ and $N3$ be the number of atoms in energy level 1, level 2 and level 3, respectively. Let P_{13} be the probability of transition of atoms from level 1 to level 3 due to pumping. The atoms decay from level 3 to level 2 with a probability P_{32}. The atoms then come back to level 1 from level 2 with a transition probability P_{21}. While the atoms move from level 2 to level 1, laser is produced due to stimulated emission. It may be more appropriate to also consider the decay of few atoms directly from level 3 to level 1, with a transition probability P_{31}, as shown in Figure 8.7.

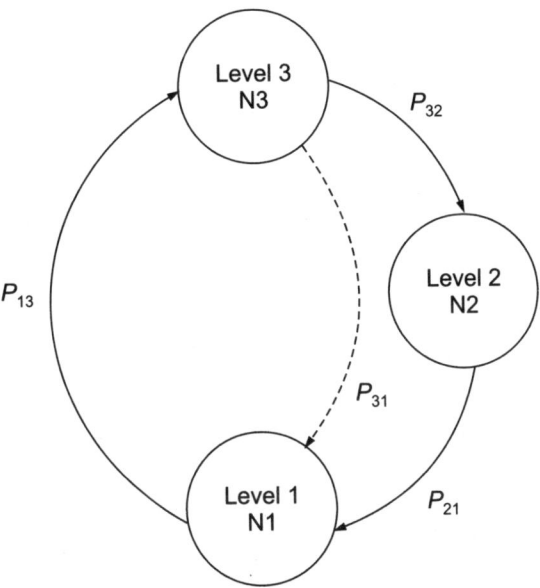

FIGURE 8.7 Transitions in the three level laser process.

Let $N1_0$, $N2_0$ and $N3_0$ be the initial number of atoms in level 1, level 2 and level 3, respectively. Now, the discrete model of the process can be formulated as follows:

$$\left. \begin{array}{l} N1_1 = P_{31}N3_0 + P_{21}N2_0 \\ N2_1 = P_{32}N3_0 \\ N3_1 = P_{13}N1_0 \end{array} \right\}$$

As discussed earlier, the mathematician must only consider the incoming lines for each level for obtaining the model. In a generalised way, the above model can be rewritten as

$$\left. \begin{array}{l} N1_{t+1} = P_{31}N3_t + P_{21}N2_t \\ N2_{t+1} = P_{32}N3_t \\ N3_{t+1} = P_{13}N1_t \end{array} \right\}$$

EXERCISES

1. Develop a model based on ordinary differential equations for the three-level lasing process and four-level lasing process.
2. For the four-level lasing process, as shown in the following figure, obtain the discrete equations governing the process.

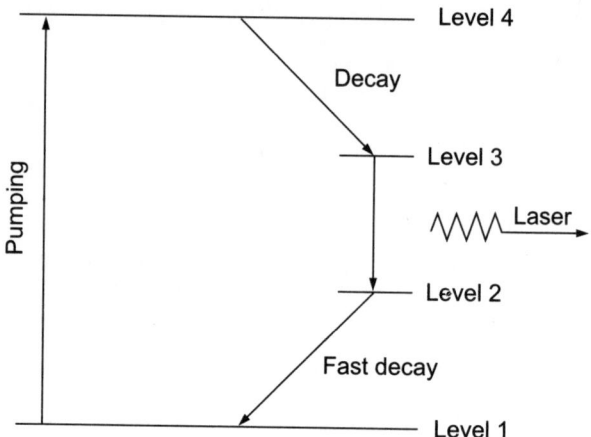

3. Develop a mathematical model of a civil war by considering three different populations:
   (a) Government
   (b) Insurgents
   (c) Civilians

4. Using the Lotka–Volterra model, analyse the conditions at which both the species are stable and their populations are at a steady state.

5. Non-dimensionalise the Lotka–Volterra predator-prey model.

6. Develop a mathematical model for the interaction of a predator with two competing prey species.

CHAPTER 9
Mathematical Modelling using the Calculus of Variations

By the method of differential calculus, the function $f(x_1, x_2, \ldots, x_n)$ has a local maximum at the point $x_{1m}, x_{2m}, \ldots, x_{nm}$, if all the first order partial derivatives vanish at this point and if the matrix of all the second order partial derivatives at this point is negative definite.

In the method of calculus of variations, we find functions $f_1(x_1, x_2, \ldots, x_n)$, $f_2(x_1, x_2, \ldots, x_n), \ldots, f_n(x_1, x_2, \ldots, x_n)$, for which a function of these functions is maximum or minimum. The mathematical problems in which we find functions for a maximum or a minimum are called *variational problems*.

Such variational problems can be used to effectively obtain the mathematical model of systems. The common examples of variational problems are as follows:

1. Finding a line of minimum length connecting two given points. (In this problem, the functional involved is the length of the line and the required line is the line segment connecting the given two points.)
2. Finding the curve connecting two given points along which a mass M slides down under gravitational force in the shortest time. (In this problem, the functional involved is the time during which the mass moves along the length of the curve and the curve minimising the functional is called the *brachistochrone*.)
3. Finding a closed curve of given length bounding a maximum area. (In this case, the functional involved is the area of the closed curve.)

A *functional* is a variable quantity whose values are determined by the choice of one or more functions. If R is a set of functions, then the functional $I(f)$ is said to be defined on the set R if every element $f \in R$ is associated with a certain number $I(f)$.

The simplest variational problem is to find a family of functions $y = y(x)$ that satisfies the boundary conditions $y(x_1) = y_1$ and $y(x_2) = y_2$ and extremizes a functional of the form

$$I(y) = \int_{x_1=a}^{x_2=b} F(x, y, y') dx$$

where, $y \in C^{(1)}$. Here, $C^{(1)}$ denotes the space of continuously differentiable functions f on $[x_1, x_2]$ with norm $\|f\|_1 = \max_{x \in [x_1, x_2]} |f(x)| + \max_{x \in [x_1, x_2]} |f'(x)|$.
$F(x, y, y')$ is a continuous function having continuous partial derivatives with respect to all variables up to the second order.

Consider the functional $I(y)$; the increment of variation δy of the argument y of the functional $I(y)$ is defined as the difference between y and $\tilde{y}$, where, $\tilde{y} \in C^{(1)}$.

$$\delta y = \tilde{y} - y$$

The quantity $\Delta I = \Delta I(\delta y) = I(y + \delta y) - I(y)$ is called the increment of a functional corresponding to the increment δy.

If we are interested in the functionals $I(y)$ defined on a set of n-times differentiable functions $y(x)$, the variation of the argument $\delta y = \tilde{y}(x) - y(x)$ is a function of x which can be differentiated n times

$$(\delta y)' = \tilde{y}' - y' = \delta y'$$
$$(\delta y)'' = \tilde{y}'' - y'' = \delta y''$$
$$\vdots$$
$$(\delta y)^{(n)} = \tilde{y}^{(n)} - y^{(n)} = \delta y^{(n)}$$

Hence, if a function $y(x)$ receives the increment δy, then its kth derivative $y^{(k)}(x)$ receives the increment $(\delta y)^{(k)} = \delta y^{(k)}$, $k = 1, 2, ..., n$.

Now, consider the functional $I(y) = \int_a^b F(x, y, y') dx$. Let us find the variation of the functional. Let $\delta y = h(x)$ be the increment of the function $y(x)$. Then, $\delta y' = h'$. Applying Taylor's formula to the difference of the integrands, we get

$$\Delta I = \int_a^b [F(x, y+h, y'+h') - F(x, y, y')] dx$$

$$= \int_a^b [F_y(x, y, y')h + F_{y'}(x, y, y')h'] dx$$

$$+ \frac{1}{2} \int_a^b \left[F_{yy}(x, y^*, y^{*\prime})h^2 + 2F_{yy'}(x, y^*, y^{*\prime})h\tilde{h}' + F_{y'y'}(x, y^*, y^{*\prime})h'^2 \right] dx$$

(9.1)

where, $y^* = y + h^* \in C^{(1)}$.

$$\Delta I = \int_a^b (F_y h + F_{y'} h') dx + I_2(h)$$

where, $I_2(h)$ is the second integral in Eq. (9.1) and

$$\frac{\partial F}{\partial y} = F_y$$

$$\frac{\partial F}{\partial y'} = F_{y'}$$

$$\frac{\partial^2 F}{\partial y \partial y'} = F_{yy'}$$

$$\vdots$$

The first integral is additive and continuous with respect to $\delta y = h$, and it can be shown that the second integral tends to zero as $\|h\|_1 \to 0$. Hence, we can write

$$\delta I = \int_a^b (F_y h + F_{y'} h') dx$$

Further, we can show that the variation of the functional

$$I(y_1, ..., y_n) = \int_a^b F(x, y_1, y_1', ..., y_n, y_n') dx$$

dependent on n functions $y_1, ..., y_n$ has the form

$$\delta I = \int_a^b \sum_{i=1}^n (F_{y_i} h_i + F_{y_i'} h_i') dx$$

Also, the variation of the functional

$$I(y) = \int_a^b F(x, y_1, y_1', ..., y^{(k)}) dx$$

dependent on the function and its derivatives up to the kth order, has the form

$$\delta I = \int_a^b (F_y h + F_{y'} h' + ... + F_{y^{(k)}} h^{(k)}) dx$$

The functional $I(y)$ reaches an extremum for $y = y_0(x)$ if there is a ε-neighbourhood of the point $y = y_0(x)$

$$P(y_0, \varepsilon) = \{y \in C^{(1)}, \|y - y_0\|_1 < \varepsilon\}$$

in which the increment $\Delta I = I(y) - I(y_0)$ does not change its sign; and if $\Delta I > 0$, then $I(y)$ reaches a minimum for $y = y_0$; and if $\Delta I < 0$, then it reaches a maximum.

If a functional $I(y)$ reaches an extremum for $y = y_0$, then it vanishes for $y = y_0$, i.e.,

$$\delta I \big|_{y=y_0} = 0$$

This is a necessary condition for the existence of an extremum of a functional.

9.1 EULER'S EQUATION

For the functional

$$I(y) = \int_a^b F(x, y_1, y_1') \, dx \tag{9.2}$$

considered on a set of functions $y(x)$ satisfying the boundary conditions

$$y(a) = A, \; y(b) = B \tag{9.3}$$

the variational problem is stated as:

"Among all functions $y(x)$ with the boundary conditions shown in Eq. (9.3), find the conditions which extremise the functional given in Eq. (9.2).

If the function $y = y(x)$ satisfies the conditions of Eq. (9.3) and extremises the functional given in Eq. (9.2), then it can be shown that y is a solution of the Euler's equation.

$$F_y - \frac{d}{dx} F_{y'} = 0 \tag{9.4}$$

There are three cases, when Eq. (9.4) allows the reduction of order.

Case I

The function $F(x, y, y')$ does not contain y explicitly, i.e., $F(x, y, y') = F(x, y')$. In this case, $F_y = 0$ and Euler's equation takes the following form:

$$\frac{d}{dx} F_{y'} = 0 \; \text{ or } \; F_{y'} = C \tag{9.5}$$

where C is a constant. Hence, we have a first order differential equation not containing y explicitly.

Case II

The function $F(x, y, y')$ does not contain x explicitly, i.e., $F(x, y, y') = F(y, y')$. Now, the Euler's equation takes the following form:

$$\frac{d}{dx}(F - y'F_{y'}) = 0 \; \text{ or } \; F - y'F_{y'} = C \tag{9.6}$$

Hence, we obtain a first order differential equation not containing x explicitly.

Case III

If the function $F(x, y, y')$ is independent of y', i.e., $F = F(x, y)$, then the Euler's equation takes the following form:

$$F_y = 0 \tag{9.7}$$

which is an equation defining a single or a family of curves.

EXAMPLE 9.1 Consider a bead of mass M which moves from point $A(1, 0)$ to the point $B(2, 1)$ with velocity $v = x$. Find the curve for which the motion time is minimum.

Solution: The time taken by the bead to cover the arc length of the curve $y = y(x)$ is given by the integral

$$t(y) = \int_1^2 \frac{\sqrt{1+y'^2}}{x} dx$$

which represents a functional in which the curves $y(x)$ satisfy the conditions $y(1) = 0$ and $y(2) = 1$.

Here, $F(x, y, y') = \dfrac{\sqrt{1+y'^2}}{x}$, is independent of y.

Hence, the Euler's equation is

$$F_{y'} = C, \text{ where } C \text{ is a constant, Let } C_1 = 1/C$$

$$F_{y'} = \frac{y'}{x\sqrt{1+y'^2}} = C = \frac{1}{C_1}$$

and

$$y' = \pm \frac{x}{\sqrt{C_1^2 - x^2}}$$

Integrating, we get

$$y = C_2 \pm \sqrt{C_1^2 - x^2}, \text{ where } C_2 \text{ is a constant.}$$

or

$$y - C_2 = \pm\sqrt{C_1^2 - x^2}$$

or

$$(y - C_2)^2 = C_1^2 - x^2$$

or

$$(y - C_2)^2 + x^2 = C_1^2$$

By applying the boundary conditions, we can obtain the values of C_1 and C_2.
When $x = 1$, $y = 0$ and when $x = 2$, $y = 1$.
Hence,

$$(0 - C_2)^2 + 1^2 = C_1^2 \tag{i}$$

$$C_2^2 + 1 = C_1^2 \tag{ii}$$

Subtracting Eq. (ii) from Eq. (i), we get

$$(1-C_2)^2 - C_2^2 + 4 - 1 = 0$$

$$1 + C_2^2 - 2C_2 - C_2^2 + 3 = 0$$

$$4 - 2C_2 = 0$$

$$\Rightarrow C_2 = 2$$

From Eq. (i),

$$4 + 1 = C_1^2$$

Therefore, we get the required equation as:

$$(y-2)^2 + x^2 = 5$$

EXAMPLE 9.2 Among the curves connecting two points (x_1, y_1) and (x_2, y_2), find the curve whose rotation about the axis of abscissas generates a surface of minimum area.

Solution: The area of a surface of revolution about the x-axis is given by the functional

$$I(y) = 2\pi \int_{x_1}^{x_2} y\sqrt{1+y'^2}\ dx$$

which represents the permissible curves satisfying the conditions $y(x_1) = y_1$ and $y(x_2) = y_2$. The functional involved in this problem does not contain x, and hence, the Euler's formula can be written as

$$F - y'F_{y'} = C$$

$$y\sqrt{1+y'^2} - \frac{yy'^2}{\sqrt{1+y'^2}} = C$$

$$y(1+y'^2) - yy'^2 = C\sqrt{1+y'^2}$$

$$y(1+y'^2 - y'^2) = C\sqrt{1+y'^2}$$

$$y = C\sqrt{1+y'^2}$$

Put $y' = \sinh t$

$$\frac{y}{\sqrt{1+y'^2}} = \frac{y}{\cosh t} = C$$

$$\therefore \qquad y = C \cosh t$$

$$dy = C \sinh t.\, dt$$

$$dx = \frac{dy}{y'} = \frac{C \sinh t\, dt}{\sinh t} = C\, dt$$

$$dx = C\, dt$$
$$\therefore \quad x = Ct + C_1$$
$$t = \frac{x - C_1}{C}$$

$y = C \cosh t = C \cosh\left(\dfrac{x - C_1}{C}\right)$, which represents two-parameter family of catenaries.

EXAMPLE 9.3 Find the shortest smooth plane curve joining two points in the plane.

Solution: Consider two distinct points (x_1, y_1) and (x_2, y_2) and $y = f(x)$ be the equation of any curve passing through (x_1, y_1) and (x_2, y_2). The length of the curve is given by

$$L(y) = \int_{x_1}^{x_2} \sqrt{1 + y'^2}\, dx \qquad (i)$$

Finding the plane curve whose length is the shortest is to find the function $y(x)$, which minimises the functional given in Eq. (i).

Euler's equation is given by

$$\frac{\partial f}{\partial y} - \frac{d}{dx}\frac{\partial f}{\partial y'} = 0$$

where,
$$f = \sqrt{1 + y'^2}$$

$$\frac{\partial f}{\partial y} = 0,\ \frac{\partial f}{\partial y'} = \frac{2y'}{2\sqrt{1 + y'^2}}$$

$$\frac{d}{dx}\frac{y'}{\sqrt{1 + y'^2}} = 0$$

$$\frac{y'^2}{\sqrt{1 + y'^2}} = C,\ \text{a constant}$$

$$y'^2 = C^2(1 + y'^2)$$
$$y'^2(1 - C^2) = C^2$$

$$y'^2 = \frac{C^2}{1 - C^2} = k\ \text{a constant}$$

Hence, $y = Cx + d$ is a straight line.

EXAMPLE 9.4 Determine if the plane curve down which a particle will slide without friction from point $A(x_1, y_1)$ to $B(x_2, y_2)$ in the shortest time.

Solution: Assume the positive direction of the y-axis is vertically downwards and let $x_1 < x_2$. Let $P(x, y)$ be the position of the particle at any time t on the curve C.

The speed of the particle v sliding along any curve is given by $v = \sqrt{2gy}$. Assume initial speed is 0 and $x_1 = 0$, $y_1 = 0$.

$$\frac{ds}{dt} = \sqrt{2gy}$$

Time taken for the particle moving under gravity from $A(0, 0)$ to $B(x_2, y_2)$ is

$$t(y(x)) = \int \frac{ds}{\sqrt{2gy}} = \frac{1}{\sqrt{2g}} \int_0^{x_2} \frac{\sqrt{1+y'^2}}{\sqrt{y}} \, dx$$

Subject to $y(0) = 0$, $y(x_2) = y_2$.

$$f = \frac{\sqrt{1+y'^2}}{\sqrt{y}}$$

Euler's equation is

$$\frac{d}{dx}\left(f - y'\frac{\partial f}{\partial y'}\right) = 0$$

$$\frac{\sqrt{1+y'^2}}{\sqrt{y}} - \frac{y'^2}{\sqrt{y}\sqrt{1+y'^2}} = k_1$$

$$(1+y'^2) - y'^2 = k_1 \sqrt{y}\sqrt{1+y'^2}$$

$$\frac{1}{k_1} = \sqrt{y}\sqrt{1+y'^2}$$

$$y(1+y'^2) = k_2 \text{ a constant}$$

Put $y' = \cot \theta$

$$y = \frac{k_2}{1+y'^2} = \frac{k_2}{1+\cot^2 \theta} = k_2 \sin^2 \theta = \frac{k_2}{2}(1 - \cot 2\theta)$$

$$dy = \frac{k_2}{2} 2 \sin 2\theta \, d\theta$$

$$\frac{dy'}{dx} = y'$$

$$dx = \frac{k_2 \sin 2\theta d\theta}{\cot \theta} = \frac{k_2 \, 2\sin^2\theta \cos\theta \, d\theta}{\cos\theta}$$

$$= 2k_2 \sin^2 \theta d\theta$$
$$= k_2(1 - \cos 2\theta)d\theta$$

$$x = \frac{k_2}{2}(2\theta - \sin 2\theta) + k_3$$

$$y(0) = 0 \implies k_3 = 0$$

$$\therefore \quad x = \frac{k_2}{2}(2\theta - \sin 2\theta)$$

Hence, we get

$$x = \frac{k_2}{2}(2\theta - \sin 2\theta) = \frac{k_2}{2}(\phi - \sin \phi)$$

$$y = \frac{k_2}{2}(1 - \cos 2\theta) = \frac{k_2}{2}(1 - \cos \phi), \quad \text{where } \phi = 2\theta.$$

It represents a one-parameter family of cycloids, with $k_2/2$ as the radius of the rolling circle.

EXAMPLE 9.5 Find the extremals of the functional

$$I(y) = \int_0^1 (x - y)^2 dy$$

with $y(0) = 1$, $y(1) = 2$.

Solution: Here, y' is not present in the functional, and hence, the Euler's formula is given as

$$F_y = 0$$
$$F_y = -2(x - y) = 0 \qquad \text{(i)}$$

However, Eq. (i) does not satisfy the boundary conditions and this problem does not have any solution.

9.2 VARIATIONAL PROBLEM FOR A FUNCTIONAL DEPENDING ON n FUNCTIONS

Suppose there is a functional

$$I(y_1, y_2, \ldots, y_n) = \int_a^b F(x, y_1, y_1', \ldots, y_n, y_n') dx \qquad (9.8)$$

which is dependent on n functions $y_1, y_2, \ldots, y_n$ and satisfies the boundary conditions

$$y_i(a) = A_i, \, y_i(b) = B_i, \quad i = 1, 2, \ldots, n \qquad (9.9)$$

Then, a system of linearly independent functions $y_1(x), y_2(x), \ldots, y_n(x)$ satisfies conditions given in Eq. (9.9) and extremizing functional Eq. (9.8) is a solution of the system of Euler's differential equations.

$$F_{y_i} - \frac{d}{dx} F'_{y_i} = 0, \quad i = 1, 2, \ldots, n$$

9.3 VARIATIONAL PROBLEM FOR FUNCTIONAL DEPENDENT ON HIGHER ORDER DERIVATIVES

Consider the following functional:

$$I(y) = \int_a^b F(x, y, y', y'', \ldots, y^{(k)}) \, dx \tag{9.10}$$

Suppose it is given on a set of functions and satisfies boundary conditions:

$$y(a) = A, \; y'(a) = A_1, \; y''(a) = A_2, \ldots, y^{(k-1)}(a) = A_{k-1}$$
$$y(b) = B, \; y'(b) = B_1, \; y''(b) = B_2, \ldots, y^{(k-1)}(b) = B_{k-1} \tag{9.11}$$

If the function $y(x)$ satisfies the boundary conditions given in Eq. (9.11) and extremizes the functional given in Eq. (9.10), then it is a solution of the Euler–Poisson differential equation.

$$F_y - \frac{d}{dx} F_{y'} + \frac{d^2}{dx^2} F_{y''} - \ldots + (-1)^k \frac{d^k}{dx^k} F_{y^{(k)}} = 0$$

EXAMPLE 9.6 Find the extremals of the functional $I = \int_1^2 (y'^2 + z^2 + z'^2) \, dx$, given $y(1) = 1, \; y(2) = 2, \; z(1) = 0, \; z(2) = 2$.

Solution: Here, $y'^2 + z^2 + z'^2$.

$$F_y - \frac{d}{dx} F_{y'} = 0 \Rightarrow y'' = 0 \Rightarrow y = C_1 x + C_2$$

$$F_z - \frac{d}{dx} F_{z'} = 0 \Rightarrow 2z - \frac{d}{dx} 2z' = 0 \Rightarrow z'' - z = 0 \Rightarrow z = C_2 e^x + C_4 e^{-x}$$

$$y(1) = 1 \Rightarrow 1 = C_1 + C_2$$
$$y(2) = 2 \Rightarrow 2 = 2C_1 + C_2$$

On solving, we get

$$C_1 = 1, \; C_2 = 0$$

$\therefore \quad y = x$

$$z(1) = 0 \Rightarrow 0 = C_3 e + C_4 e^{-1}$$
$$z(2) = 1 \Rightarrow 1 = C_3 e^2 + C_4 e^{-2}$$

On solving, we get

$$C_3 = \frac{1}{e^2-1}, \quad C_4 = \frac{-e^2}{e^2-1}$$

$$\therefore \quad z = \frac{1}{e^2-1}e^x - \frac{e^2}{e^2-1}e^{-x}$$

Hence, the extemals are $y = x$ and $z = \frac{1}{e^2-1}e^x - \frac{e^2}{e^2-1}e^{-x}$.

EXAMPLE 9.7 Find the extremals of

$$I = \int_0^1 (360x^2 y - y''^2) dx$$

given $y(0) = 0$, $y'(0) = 1$, $y(1) = 0$, $y'(1) = 2.5$.

Solution:

$$F_y - \frac{d}{dx}F_{y'} + \frac{d^2}{dx^2}F_{y''} = 0, \quad \text{where } F = 360x^2 y - y''^2$$

$$\therefore \quad 360x^2 + \frac{d^2}{dx^2}(-2y'') = 0$$

$$y^{iv}(x) = 180x^2$$

$$y(x) = \frac{x^6}{2} + C_1 x^3 + C_2 x^2 + C_3 x + C_4$$

Using conditions, we get

$$C_1 = 3/2, \quad C_2 = -3, \quad C_3 = 1, \quad C_4 = 0$$

$$\therefore \quad y = \frac{x^4}{2} + \frac{3}{2}x^3 - 3x^2 + x$$

9.4 APPLICATIONS OF CALCULUS OF VARIATIONS

In several fields of science and engineering, the calculus of variations acts as a powerful modelling tool. For example, the calculus of variation can be used to obtain the models of certain systems using the Hamilton's principle, which states that any dynamical system will always follow the trajectory that minimises the action integral

$$A = \int_{t_1}^{t_2} L\,dt = \int_{t_1}^{t_2} (T - V)\,dt$$

where, L is the Lagrangian, which is the difference between the kinetic energy T and the potential energy V of the system. In nature, systems behave in such a way that the difference between the kinetic energy and the potential energy is minimised.

In several systems, the kinetic energy depends only on the velocity of the particle and the potential energy depends only on the position of the particle. Hence, the Lagrangian L is a function of time t, position y and velocity (y'). Hence, the action integral takes the form

$$A = \int_{t_1}^{t_2} L(t, y, y') \, dt$$

For example, consider the trajectory of a projectile of mass m. Let the air friction be negligible and the projectile is launched at time $t_1 = 0$ from an initial position $(x(0), y(0)) = (0, h)$ with an initial velocity v_0 and an angle θ above the horizontal plane. Hence, the components of the velocity vector will be

$$(v_{0x}, v_{0y}) = |v_0|(\cos\theta, \sin\theta)$$

The action integral of the projectile motion is given by

$$A = \int_{t_1}^{t_2} L(t, x, y, x', y') \, dt$$

$$= \int_{t_1}^{t_2} (T - V) \, dt$$

The kinetic energy of the particle is $T = \frac{1}{2} m |v|^2$ and the potential energy is $V = mgy$. Since the magnitude $|v|^2 = (x')^2 + (y')^2$, the action integral is written as

$$A = \int_{t_1}^{t_2} m\left(\frac{1}{2}(x')^2 + \frac{1}{2}(y')^2 - gy\right) dt$$

and the Lagrangian $L = m\left(\frac{1}{2}(x')^2 + \frac{1}{2}(y')^2 - gy\right)$

Let $f = \dfrac{L}{m}$, and using Euler–Lagrange equations, we get

$$0 = \frac{\partial f}{\partial x} - \frac{d}{dt}\left(\frac{\partial f}{\partial x'}\right) = -\frac{d}{dt}(x') = -x''$$

and

$$x' = v_{0x} = |v_0| \cos\theta$$

On integration, we get
$$x = v_{0x}t + a$$
Using $x(0) = 0$, we get $a = 0$, and hence,
$$x = v_{0x}t = |\mathbf{v}_0|t \cos \theta$$
The second Euler–Lagrange equation is
$$0 = \frac{\partial f}{\partial y} - \frac{d}{dt}\left(\frac{\partial f}{\partial y'}\right) = -g - \frac{dy'}{dt} = -g - y''$$
or
$$y'' = -g$$
Integrating, we get
$$y' = -gt + b$$
At $t = 0$, $y' = |\mathbf{v}_0| \sin \theta$. Hence,
$$y' = -gt + |\mathbf{v}_0| \sin \theta$$
Integrating again, we get
$$y = \frac{-gt^2}{2} + |\mathbf{v}_0|t \sin \theta + c$$

Using $y(0) = h$, we get $c = h$. Thus, the trajectory of the projectile is given by
$$x = v_{0x}t = |\mathbf{v}_0|t \cos \theta$$
and
$$y = \frac{-gt^2}{2} + |\mathbf{v}_0|t \sin \theta + h$$

EXERCISES

1. Find the curve joining the points (x_1, y_1) and (x_2, y_2), which yields a surface of revolution of minimum area when revolved about the x-axis.

2. Prove that if a particle moves from a point P_1 to a point P_2 in a time interval $t_1 \leq t \leq t_2$, then the actual path it follows is the one for which the action assumes a stationary value.

3. Derive the equations of vibrations of a spring and a membrane using the calculus of variations.

4. Find the curve connecting given points A and B, which is traversed by a particle sliding from A to B in the shortest time, assuming that friction and air resistance are negligible.

CHAPTER 10

Stability Theory

Any mathematical model of a real system should have its overall stability corresponding to the stability of the actual system in the real world. The concept of fixed points and equilibrium comes into picture, since every real system undergoes small perturbations at all times through background noise.

10.1 FIXED POINTS OF A SYSTEM

Consider an autonomous dynamic system represented by

$$\frac{dx}{dt} = f(x)$$

The fixed points of the system can be visualised by graphically plotting dx/dt as a function of x. For example, consider the system

$$\dot{x} = \frac{dx}{dt} = r + x^2 \qquad (10.1)$$

where, r is a parameter of the system. If $r < 0$, the plot of dx/dt as function of x appears as shown in Figure 10.1.

Now, it is seen that there are two points A and B at which the rate of change of x is zero. The direction of flow (represented by the arrow head) is to the right when $\dot{x} > 0$ and to the left when $\dot{x} < 0$. At points where $\dot{x} = 0$, there is no flow and these special points are called *fixed points* (or sometimes *critical* or *singular points*). In Figure 10.1, point A is called *stable fixed point*, since all the flow is towards point A. Point B is called *unstable fixed point*, since all the flow is away from it.

Now, consider the case where $r = 0$, in Eq. (10.1). In this case, the plot of $\dot{x}$ as a function of x will appear as shown in Figure 10.2.

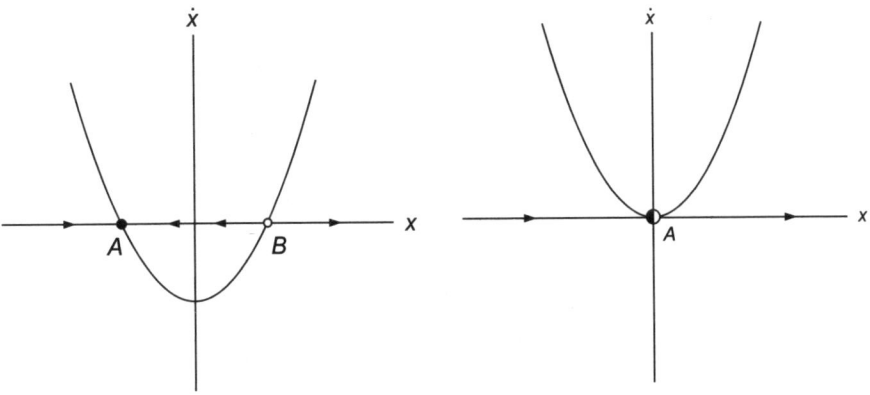

FIGURE 10.1 $r < 0$. **FIGURE 10.2** $r = 0$.

Here, we have one fixed point A. The flow in the right half plane is away from A and the flow in the left half plane is towards A. In this case, point A is called *half stable* or *semi stable fixed point*.

The fixed points represent equilibrium or steady solutions. An equilibrium is said to be stable if all small disturbances away from it damp out in time. Hence, a stable equilibrium is represented by stable fixed points and unstable equilibrium is represented by unstable fixed points in which the disturbances grow in time.

EXAMPLE 10.1 Find all the fixed points of the system governed by the equation $\dot{x} = \sin(x)$.

Solution: To find the fixed points of the given differential equation, we set $\dot{x} = 0$ or $\sin(x) = 0$.

Hence, the fixed points of the equation are

$$x^* = n\pi, \quad n = 0, \pm 1, \pm 2, \pm 3, \ldots.$$

10.2 PHASE PLANE ANALYSIS

Consider the non-linear autonomous system represented by

$$\left. \begin{array}{l} \dfrac{dx}{dt} = f(x, y) \\ \dfrac{dy}{dt} = g(x, y) \end{array} \right\} \qquad (10.2)$$

Since both x and y are functions of t, we can write

$$\frac{dy}{dx} = \frac{g(x, y)}{f(x, y)} \qquad (10.3)$$

Equation (10.2) defines a direction field at every point except the points where both g and f vanish. Such exceptional points are called *singular points* or *critical points* or *singularities*. In other words, a singular point is a point (x_0, y_0) at which both $f(x_0, y_0) = 0$ and $g(x_0, y_0) = 0$.

Now, make the following assumptions:

1. The origin is a singular point, i.e.,

$$f(0, 0) = 0$$
$$g(0, 0) = 0$$

2. There are no other singular points near the origin.
3. f and g are analytic functions of x and y near the origin.

Following these assumptions, f and g can be expanded using Taylor's series of the form

$$f(x, y) = ax + by + P(x, y)$$
$$g(x, y) = cx + dy + Q(x, y)$$

where, a, b, c and d are constants and P and Q are higher order terms in the Taylor's series.

Hence, the linear approximation to Eq. (10.2) can be given as

$$f(x, y) = ax + by$$
$$g(x, y) = cx + dy$$

In order to determine the stability of a non-linear system in the neighbourhood of a singularity, we can investigate the solutions of the linearised system.

$$\frac{dx}{dt} = ax + by$$

$$\frac{dy}{dt} = cx + dy$$

Now, writing this system in matrix form,

$$\begin{bmatrix} \dot{x} \\ \dot{y} \end{bmatrix} = \underbrace{\begin{bmatrix} a & b \\ c & d \end{bmatrix}}_{A} \begin{bmatrix} x \\ y \end{bmatrix}$$

and finding the determinant of A and equating to zero, we get

$$\begin{vmatrix} a - \lambda & b \\ c & d - \lambda \end{vmatrix} = 0$$

which yields the characteristic equation for λ as

$$\lambda^2 - (a + d)\lambda + (ad - bc) = 0$$

or
$$\lambda = \frac{a+d}{2} \pm \frac{1}{2}\sqrt{(a-d)^2 + 4bc}$$

Now, consider $\Delta = (a-d)^2 + 4bc$. Based on the values of Δ, the singularities are classified.

10.2.1 Classification of Singularities

1. Node: A singularity is called a node if all the solution curves go through the singularity. For example, consider the system

$$\left.\begin{array}{l}\dfrac{dx}{dt} = ax \\ \dfrac{dy}{dt} = ay\end{array}\right\} \quad (10.4)$$

whose solutions are

$$\left.\begin{array}{l}x(t) = C_1 e^{at} \\ y(t) = C_2 e^{at}\end{array}\right\} \quad (10.5)$$

where, C_1 and C_2 are constants of integration. Now, eliminating t from Eq. (10.5), we get

$$C_2 x = C_1 y$$

which is a straight line through the origin in the phase plane for any value of C_1 and C_2, not both zero. Such a singularity is called a *node*.

Now, consider the case where $a < 0$ in Eq. (10.5). In this case, as $t \to \infty$, both x and $y \to 0$ and the node is called a *stable node*.

If $a > 0$, then as $t \to \infty$, both x and $y \to \infty$, then the node is an *unstable node*.

2. Saddle point: A singularity is called a saddle point if the solution curves in the phase plane appear as shown in Figure 10.4.

For example, consider the following system:

$$\left.\begin{array}{l}\dfrac{dx}{dt} = ax \\ \dfrac{dy}{dt} = by\end{array}\right\} \quad (10.6)$$

Its analytical solution is

$$\left.\begin{array}{l}x(t) = C_1 e^{at} \\ y(t) = C_2 e^{bt}\end{array}\right\} \quad (10.7)$$

Now, eliminating t, we get

$$C_4 y^a = C_3 x^b$$

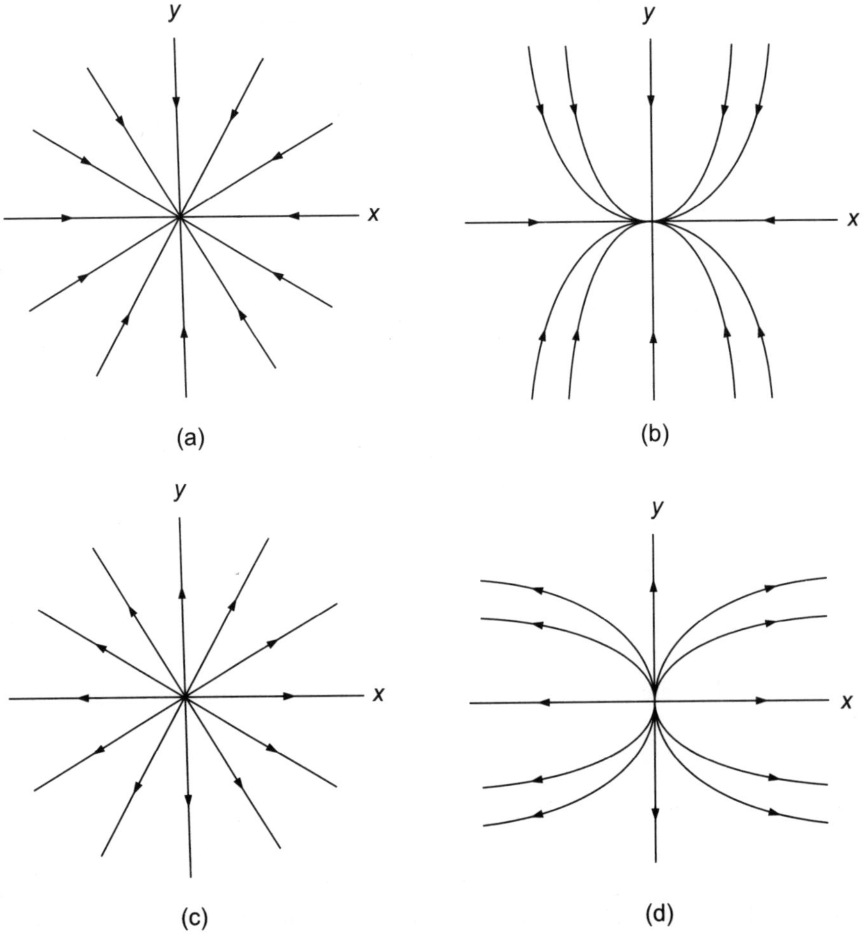

FIGURE 10.3 Node [stable (a, b) and unstable (c, d)].

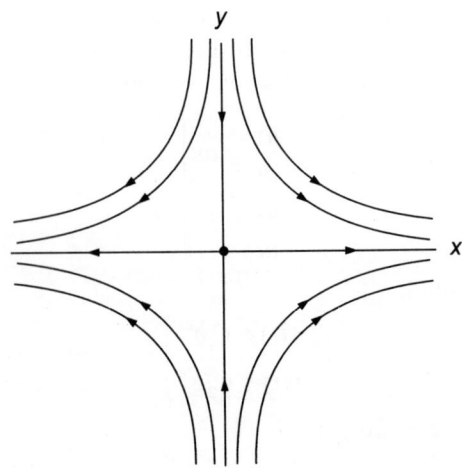

FIGURE 10.4 Saddle point.

If a and b have the same sign, then the singularity is again a node, since all the curves will go through the origin.

However, if a and b have different signs, then when $x \to 0$, $y \to \infty$ and when $x \to \infty$, $y \to 0$. Such a singularity is called *saddle point*.

When $C_3 = 0$ or $C_4 = 0$, we get $x = 0$ and $y = 0$ and these are the only ones which enter the singularity.

3. Centre: A singularity is called a centre (Figure 10.5) if it is surrounded by a family of closed paths and is not approached by any path as $t \to \infty$ or $t \to 0$.

For example, consider the following system:

$$\left. \begin{aligned} \frac{dx}{dt} &= -y \\ \frac{dy}{dt} &= x \end{aligned} \right\} \quad (10.8)$$

Its general solution is

$$\left. \begin{aligned} x(t) &= -C_1 \sin t + C_2 \cos t \\ y(t) &= C_1 \cos t + C_2 \sin t \end{aligned} \right\} \quad (10.9)$$

By eliminating t, we get $x^2 + y^2 = c^2$, which yields all the paths to the closed paths around the critical point $(0, 0)$. Hence, the singularity is a centre.

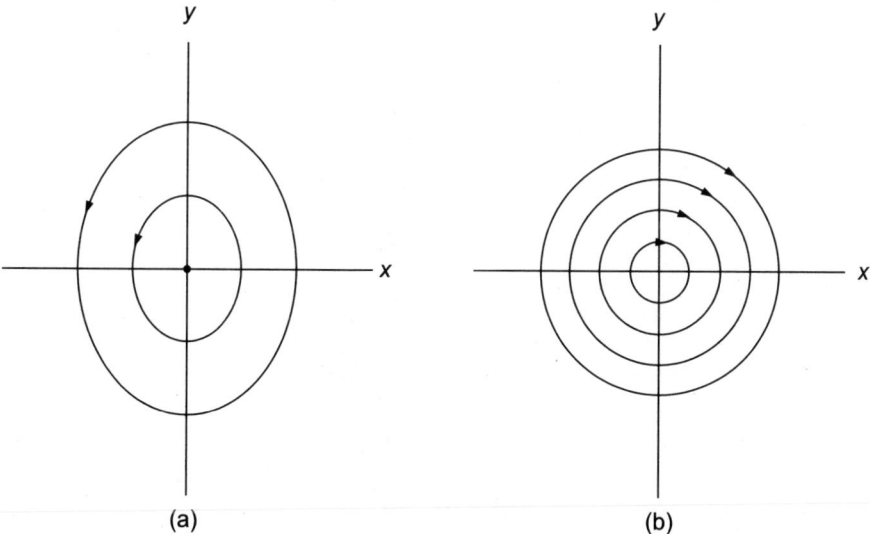

FIGURE 10.5 Centre.

4. Spiral: A spiral (Figure 10.6) or a focus is a critical point which is approached in a spiral-like manner by a family of paths that wind around it an infinite number of times as $t \to \infty$. While the paths approach the origin, they do not enter it.

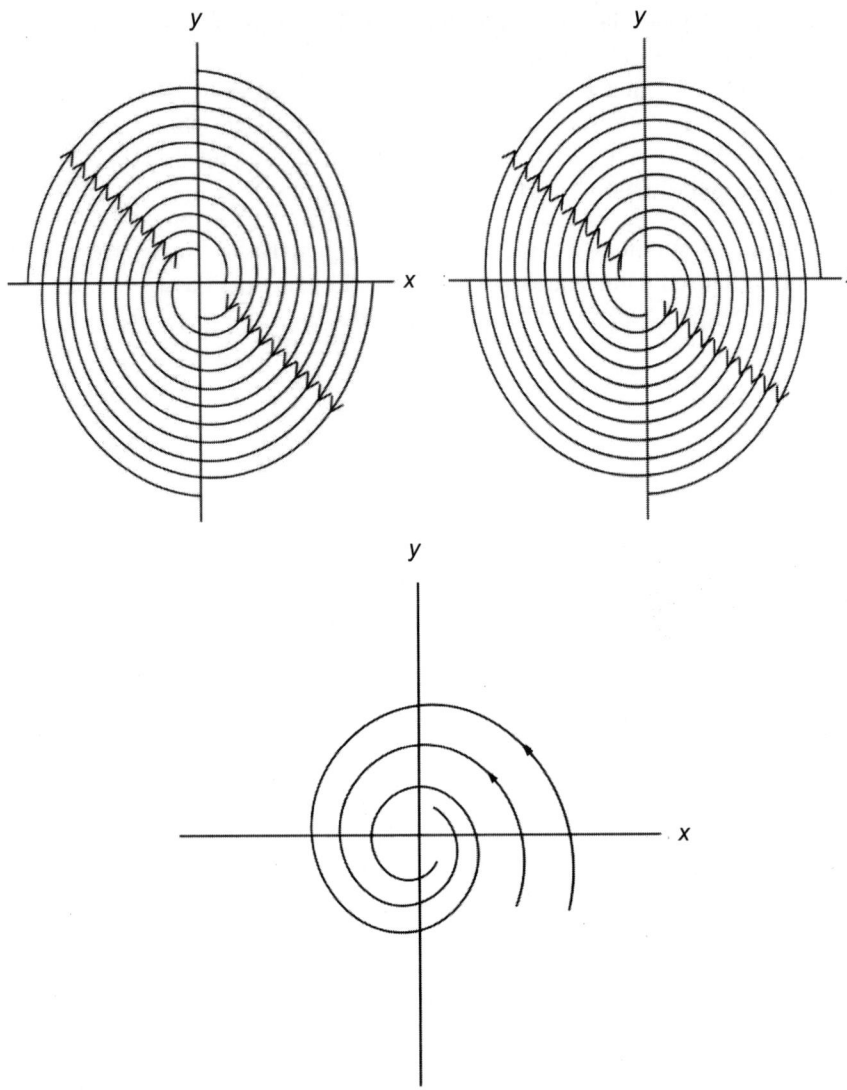

FIGURE 10.6 Spirals.

For example, consider the following system:

$$\left.\begin{aligned}\frac{dx}{dt} &= ax + by \\ \frac{dy}{dt} &= -bx + ay\end{aligned}\right\} \tag{10.10}$$

The analytical solution of the equation is

$$\left.\begin{aligned}x(t) &= e^{at} \sin bt \\ y(t) &= e^{at} \cos bt\end{aligned}\right\} \tag{10.11}$$

or
$$x^2 + y^2 = (e^{at})^2$$

or
$$\sqrt{x^2 + y^2} = e^{at}$$

If $a < 0$, the distance $\sqrt{x^2 + y^2}$ from the origin to a point moving on the solution curve will decrease steadily as the point moves around the origin. In this case, the singularity is called a *stable spiral* [Figure 10.7(a)]. If $a > 0$, the curves spiral away from the origin and the singularity is said to be an *unstable spiral* [Figure 10.7(b)].

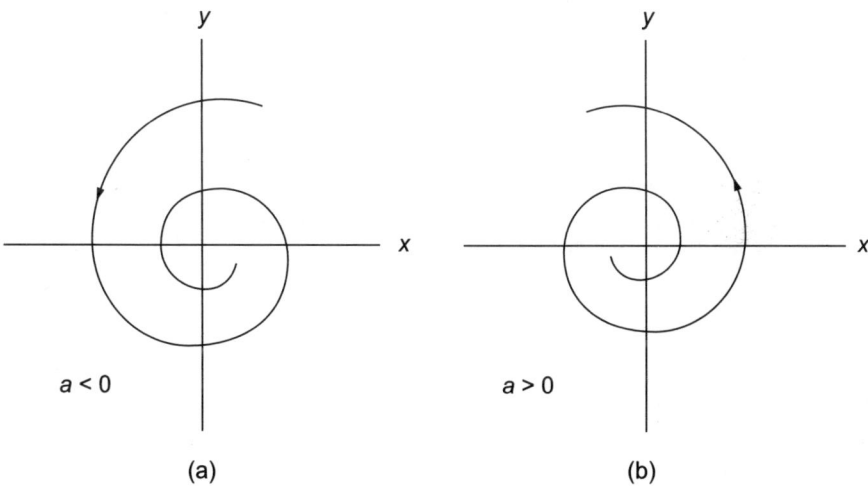

FIGURE 10.7 (a) Stable spirals and (b) Unstable spirals.

To summarise

$$\left.\begin{aligned}\frac{dx}{dt} &= ax + by \\ \frac{dy}{dt} &= cx + dy\end{aligned}\right\} \quad \text{(or)} \quad \left\{\frac{dy}{dx} = \frac{cx + dy}{ax + by}\right.$$

Let
$$\begin{cases} \Delta = (a-d)^2 + 4bc \\ p = a + d \\ q = ad - bc \neq 0 \end{cases}$$

Table 10.1 presents the classification of singularities and the conditions for stability.

To understand the concept of stability with the critical points of the system, consider a pendulum shown in Figure 10.8.

TABLE 10.1 Classification of singularities

Case	Singularity	Stability
$\Delta = 0$	Node	Stable if $p < 0$ Unstable if $p > 0$
$\Delta > 0$	Node if $q > 0$	Stable if $p < 0$ Unstable if $p > 0$
	Saddle if $q < 0$	Unstable
$\Delta < 0$	Centre if $p = 0$	Marginally stable
	Spiral if $p \neq 0$	Stable if $p < 0$ Unstable if $p > 0$

There are two steady states possible.

1. When the bob is at rest at the highest point.
2. When the bob is at rest at the lowest point.

The first state is unstable, whereas the second state is stable. The steady state of a simple physical system corresponds to an equilibrium point or a singular point in the phase plane. A small disturbance at an unstable equilibrium point leads to a larger and larger departure from this point.

A critical point is stable if all paths that get sufficiently close to the point stay close to the point.

A critical point is asymptotically stable if it is stable and there exists a circle $x^2 + y^2 = r^2$ such that every path which is inside this circle for some $t = t_0$ approaches the origin as $t \to 0$.

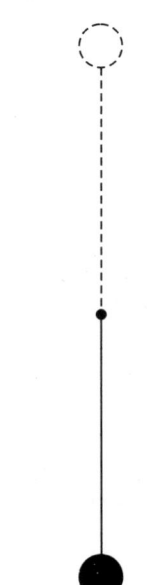

FIGURE 10.8 A pendulum.

Let us now summarse the classification of singularities for the system

$$\begin{bmatrix} \dot{x} \\ \dot{y} \end{bmatrix} = \underbrace{\begin{bmatrix} a & b \\ c & d \end{bmatrix}}_{A} \begin{bmatrix} x \\ y \end{bmatrix}$$

based on the eigenvalues λ_1, λ_2 of the A matrix:

Node: λ_1 and λ_2 are real, distinct and of the same sign.
Node: λ_1 and λ_2 are real and equal.

Stability Theory **151**

Saddle point: λ_1 and λ_2 are real, distinct and of opposite signs.
Spiral: λ_1 and λ_2 are complex conjugates but not purely imaginary.
Centre: λ_1 and λ_2 are purely imaginary.

EXAMPLE 10.2 For a mass-spring system (see the figure given below) whose equation of motion is given by

$$\frac{d^2 x}{dt^2} + x = 0 \qquad \text{(i)}$$

The characteristic equation for Eq. (i) is given by

$$m^2 + 1 = 0$$
$$m^2 = -1$$
$$m = \pm j$$

Hence, the analytical solution is

$$x(t) = A \cos t + B \sin t$$

Let $x(0) = x_0$ and $\dot{x}(0) = 0$

$$x(0) = A \cos 0 + B \sin (0)$$
$$x(0) = x_0 = A$$

Now, $\qquad \dot{x}(t) = -A \sin t + B \cos t$

Hence, $\qquad \dot{x}(0) = -A \sin(0) + B \cos(0)$

$$B = 0$$

Therefore, the solution is $x(t) = x_0 \cos t$.

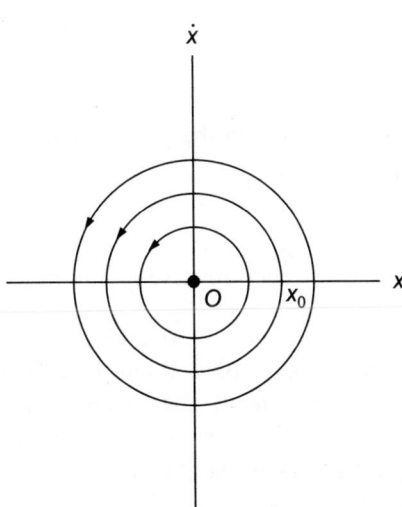

Phase portrait of unit mass-spring system.

To eliminate time,

$$x^2(t) + (\dot{x}(t))^2 = x_0^2 \cos^2 t + x_0^2 \sin^2 t$$

$$x^2 + (\dot{x})^2 = x_0^2$$

which is a series of circles of radius equal to the initial condition x_0 (see the figure Phase portrait of unit mass-spring system) in the phase plane and the system is marginally stable.

EXAMPLE 10.3 Analyze the phase trajectories of the RL-network (shown in the following figure) carrying a current i_0, and connected to a DC voltage source v without interruption of current flow.

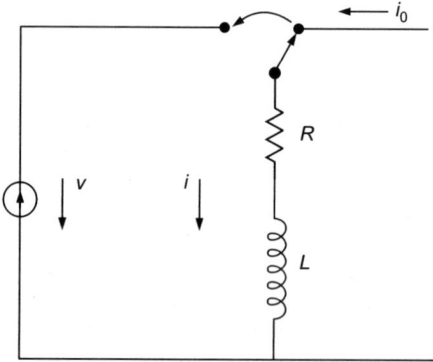

The transient response of the network will be described as

$$L\frac{di}{dt} + Ri = v$$

Let $i = x$ and $\dfrac{di}{dt} = y$.

Then, $y = (v - Rx)/L$.

The trajectory of the transient in the phase plane is a straight line passing through the points $\left(0, \dfrac{v}{L}\right)$ and $\left(\dfrac{v}{R}, 0\right)$ as shown in the figure. In the upper half plane $y = \dfrac{dx}{dt} > 0$, the representative point moves in the direction of increasing x. In the lower half plane, $y < 0$, and the representative point moves in the direction of negative values of x. Point A in figure is a stable node on the x-axis and corresponds to a steady state. Depending on the initial value of i_0 (at $t = 0$), the trajectory may have its origin at various points on the straight line. As $t \to \infty$, the point moves in the direction of the stable node A at a rate which decreases as it approaches A.

Stability Theory **153**

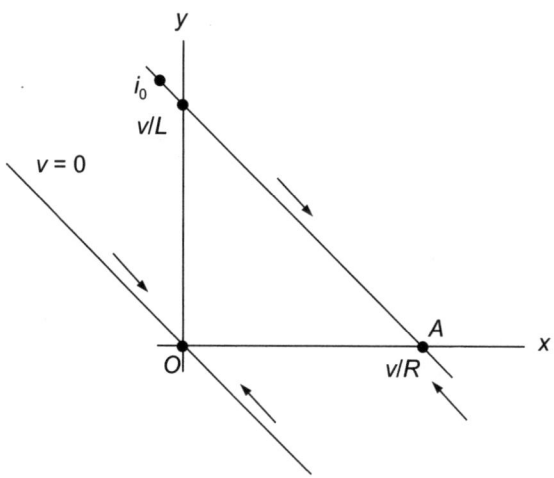

FIGURE Phase portrait of R-L network transients.

If the RL-network is short circuited ($v = 0$), the trajectory will pass through the origin of coordinates.

EXAMPLE 10.4 Construct the phase portrait of the transients in an RLC network shown in the following figure.

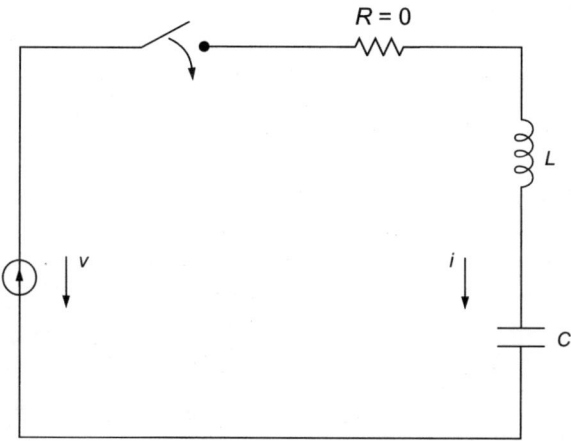

Now, consider the case where $R = 0$ and L and C are arranged in a series combination connected to a DC voltage source v. The initial values of current in the inductance and voltage in the capacitance are represented as i_0 and v_0, respectively. The transient response for this case is given by

$$\frac{d^2i}{dt^2} + i/LC = 0$$

Let $i = x$, $\dfrac{di}{dt} = y$ and $\dfrac{1}{LC} = \omega_0^2$. Then, we get

$$\frac{dy}{dx} = -\omega_0^2 \left(\frac{x}{y}\right)$$

On integration, we get

$$y^2 + \omega_0^2 x^2 = k^2$$

On the phase plane, the equation is represented by a set of ellipses whose vertical axes are equal to $2K$ and their horizontal axes are equal to $\dfrac{2K}{\omega_0}$. This phase portrait corresponds to continuous sine wave oscillations and the critical point is a centre (see the figure below).

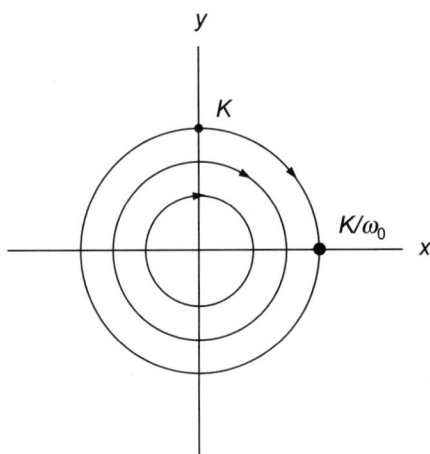

10.3 LIAPUNOV'S DIRECT METHOD

Alexander Mikhailovich Liapunov (1857–1918), a Russian Mathematician and Mechanical Engineer, proposed a simple, efficient and effective method for analysing the stability of physical systems by calculating the energy associated with the system. The idea behind Liapunov's direct method is that if the total energy of a physical system has a local minimum at a certain equilibrium point, then that point is stable.

Consider an autonomous system given by

$$\left. \begin{array}{l} \dfrac{dx}{dt} = F(x, y) \\[6pt] \dfrac{dy}{dt} = G(x, y) \end{array} \right\} \qquad (10.12)$$

Assume that this system has an isolated critical point, which we take to be at the origin (0, 0). Let $C = [x(t), y(t)]$ be a path of Eq. (10.12), and consider a function $E(x, y)$ that is continuous and has continuous first partial derivatives in a region containing this path. If a point (x, y) moves along the path in accordance with the equations $x = x(t)$ and $y = y(t)$, then $E(x, y)$ can be seen as a function of t along C. The rate of change of $E(t)$ is given by

$$\frac{dE}{dt} = \frac{\partial E}{\partial x}\frac{dx}{dt} + \frac{\partial E}{\partial y}\frac{dy}{dt}$$

$$= \frac{\partial E}{\partial x}F + \frac{\partial E}{\partial y}G$$

Suppose $E(x, y)$ is continuous and has continuous first partial derivatives in some region containing the origin. Then

1. E is said to be positive definite if E vanishes at the origin ($E(0, 0) = 0$) and if $E(x, y) > 0$ for $(x, y) \neq (0, 0)$.
2. E is negative definite if $E(x, y) < 0$ for $(x, y) \neq (0, 0)$.
3. E is positive semidefinite if $(E(0, 0) = 0)$ and $E(x, y) \geq 0$ for $(x, y) \neq (0, 0)$.
4. E is negative semidefinite if $(E(0, 0) = 0)$ and $E(x, y) \leq 0$ for $(x, y) \neq (0, 0)$.

In general, functions of the form $ax^{2m} + by^{2n}$, where a and b are positive constants and m and n are positive integers, are positive definite.

A positive definite function $E(x, y)$, whose rate of change dE/dt is negative semidefinite is called the *Liapunov function* of the system.

If $\frac{dE}{dt} \leq 0$, then E is non-increasing along the paths of Eq. (10.12) near the origin. The Liapunov functions generalise the concept of the total energy of a physical system. If the rate of change of total energy of the system is negative or zero, then it can be said that the system is stable.

If there exists a Liapunov function $E(x, y)$ for the system shown in Eq. (10.12), then the critical point $(0, 0)$ is stable and if the rate of change of this function with respect to time $[dE/dt]$ is negative definite, then the critical point $(0, 0)$ is asymptotically stable.

The Liapunov functions can be easily obtained if the associated energy of the system is known.

For example, consider the mass-spring-dashpot system whose equation of motion is given by

$$M\frac{d^2x}{dt^2} + B\frac{dx}{dt} + kx = 0$$

where $B \geq 0$ represent the viscosity damper, $k > 0$ the spring constant and M is the mass of the system. Now, decomposing the system into first order simultaneous equation,

$$\left.\begin{array}{l} \dfrac{dx}{dt} = y \\[6pt] \dfrac{dy}{dt} = -\dfrac{k}{M}x - \dfrac{B}{M}y \end{array}\right\} \quad \text{(i)}$$

and its only critical point is (0, 0). The total energy of the system can be represented as the sum of the kinetic and potential energy associated with the system.

$$\text{Kinetic energy} = \frac{1}{2}My^2$$

where, $y = \dfrac{dx}{dt}$ is the velocity.

Potential energy or the energy stored in the spring is

$$\int_0^x kx\,dx = \frac{1}{2}kx^2$$

Thus, the total energy of the system is

$$E(x, y) = \frac{1}{2}My^2 + \frac{1}{2}kx^2 \quad \text{(ii)}$$

It is clear that Eq. (ii) is positive definite and calculating dE/dt, we get

$$\frac{dE}{dt} = \frac{\partial E}{\partial x}\frac{dx}{dt} + \frac{\partial E}{\partial y}\frac{dy}{dt}$$

$$= kxy + My\left(-\frac{k}{M}x - \frac{B}{M}y\right)$$

$$= -By^2 \leq 0$$

Equation (ii) is a Liapunov function for the system shown in Eq. (i) and the critical point (0, 0) is stable. When $B > 0$, this critical point is asymptotically stable.

For any value of $B > 0$, we see that $dE/dt < 0$. Hence, the rate of change of energy of the system is negative implying that the energy of the system is decreasing. Hence, it can be said that the system is a stable system.

Now, consider the case where the energy equation of the system is not known. In such a case, the energy equation can be obtained as a quadratic function of the system. For example, consider the system

$$\left.\begin{array}{l} \dfrac{dx}{dt} = F(x, y) \\[6pt] \dfrac{dy}{dt} = G(x, y) \end{array}\right\}$$

whose energy function $E(x, y)$ is not known. The energy function can be obtained in the following manner.

$$E(x, y) = F^2(x, y) + G^2(x, y)$$

EXAMPLE 10.5 For the system

$$\frac{dx}{dt} = -x \qquad (i)$$

Construct the Liapunov function and check whether the system is stable or not.

The Liapunov function of the system is

$$E(x) = (-x)^2 = x^2 \qquad (ii)$$

It is clearly seen that Eq. (ii) is positive definite. Now, computing the rate of change of energy, we get

$$\frac{dE}{dt} = \frac{\partial E}{\partial x} \cdot \frac{dx}{dt}$$

$$= 2x \, (-x)$$

$$= -2x^2$$

Hence, $\dfrac{dE}{dt} < 0$ for all x and the system is a stable system.

10.4 POTENTIAL FUNCTIONS AND CATASTROPHES

Consider a system $\dot{x} = f(x)$. Based on the idea of the potential energy of the system, the potential function for the system can be written as

$$f(x) = -\frac{dV}{dx}$$

such that the potential energy decreases as the motion of the particle proceeds. Hence, by definition of the potential,

$$\frac{dx}{dt} = f(x) = -\frac{dV}{dx}$$

and

$$\frac{dV}{dt} = \frac{dV}{dx}\frac{dx}{dt} = -\left(\frac{dV}{dx}\right)^2 \leq 0$$

Consider the system

$$\dot{x} = x - x^3 \qquad (10.13)$$

The potential function $V(x)$ can be obtained by solving the equation $-\dfrac{dV}{dx} = x - x^3$.

$$V(x) = -\frac{1}{2}x^2 + \frac{1}{4}x^4 + C$$

By setting $C = 0$ and graphically plotting V as a function of x, we get the curve shown in Figure 10.9.

Now, compare this with the fixed points of Eq. (10.13).

$$\frac{dx}{dt} = x - x^3 = 0$$

$$x(1 - x^2) = 0$$

$$x = 0, \pm 1$$

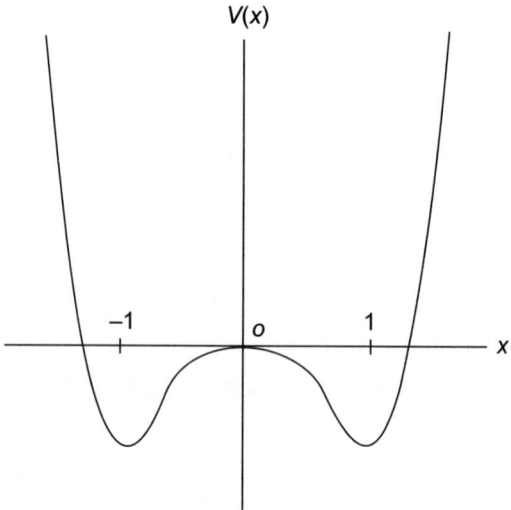

FIGURE 10.9 $V(x)$ versus x for $\dot{x} = x - x^3$.

From Figure 10.9, we see that the local maximum at $x = 0$ corresponds to an unstable equilibrium and the local minima at $x = \pm 1$ correspond to stable equilibria. Such a system is called a *bistable system*, since there are two stable equilibria.

Hence, in general, the local minima of $V(x)$ correspond to stable fixed points and local maxima of $V(x)$ correspond to unstable fixed points.

This method can be extended to n number of variables $[x_1, x_2, \ldots, x_n]$ and the equilibria of the system are the critical points of the system, where

$$\frac{\partial V}{\partial x_1} = \frac{\partial V}{\partial x_2} = \frac{\partial V}{\partial x_3} = \ldots = \frac{\partial V}{\partial x_n} = 0$$

Away from the equilibria, the function $V(x)$ decreases as t increases.

Consider an example where two biological cells of exactly the same type are influenced by the concentration $x(t)$ of a certain chemical substance at each instant of time t. Assume that the substance is entering the cells from the outside at a constant rate S and is being destroyed inside the cell at a rate proportional to its concentration. Further, assume that the cells interact with each other at a rate which is proportional to the difference in concentrations between the two cells. Now, the mathematical model of this process can be given by the following set of first order differential equations:

$$\left.\begin{array}{l}\dot{x}_1 = -ax_1 + S + D(x_1 - x_2) \\ \dot{x}_2 = -ax_2 + S + D(x_2 - x_1)\end{array}\right\}$$

The potential function $V(x_1, x_2)$ is the function which satisfies

$$\left.\begin{array}{l}\dot{x}_1 = -\dfrac{\partial V}{\partial x_1} \\ \dot{x}_2 = -\dfrac{\partial V}{\partial x_2}\end{array}\right\}$$

or
$$\dot{x} = -\text{grad } V$$

10.4.1 Catastrophe Theory

Consider the situation where the system depends on a number of parameters C_1, C_2, ..., C_k. If for each value of $C = (C_1, C_2, ..., C_k)$, the system is governed by a smooth potential function V_C, then the equilibria of the system will clearly vary with C. If C varies, it is possible for a minimum and a maximum of V_C to coalesce and disappear. Hence, if the system is in a state corresponding to the minimum which disappears (as C varies), the system has to jump or move to other minimum. This phenomenon is called *catastrophe* (Chillingworth 1976).

Hence, it is important to find the points $C_1, C_2, ..., C_n$ at which catastrophes can occur. In reality, these catastrophes would represent actual catastrophes such as the sudden breaking of an iceberg.

Now, consider that a dynamical system is governed by a smooth potential function V of n variables $x = (x_1, x_2, ..., x_n)$ with k parameters $(C_1, C_2, ..., C_k)$. The $(C_1, C_2, ..., C_k)$ space is called as the *control space* and let us define the catastrophe set k to be the set of points $C = (C_1, C_2, ..., C_k)$ in the control space for which $V_C(x)$ has some coalescent critical points, which is the set of points where we can expect catastrophic behaviour in the dynamical system.

The catastrophe set k is the set of points C such that all the partial derivatives

$$\dfrac{\partial V_C}{\partial x_1}, \dfrac{\partial V_C}{\partial x_2}, ..., \dfrac{\partial V_C}{\partial x_n}$$

and also the determinant of second derivatives.

$$\begin{vmatrix} \dfrac{\partial^2 V_C}{\partial x_1 \partial x_2} & \cdots & \dfrac{\partial^2 V_C}{\partial x_1 \partial x_n} \\ \vdots & & \vdots \\ \dfrac{\partial^2 V_C}{\partial x_n \partial x_1} & \cdots & \dfrac{\partial^2 V_C}{\partial x_n \partial x_n} \end{vmatrix}$$

vanish simultaneously for some $(x_1, x_2, \ldots, x_n)$. An equation for k in terms of $(C_1, C_2, \ldots, C_k)$ can be found by eliminating $(x_1, x_2, \ldots, x_n)$ from the equations.

For example, consider a dynamical system governed by the smooth function:

$$V_{(C_1, C_2)}(x) = x^3 + C_1 x^2 + C_2 x \tag{10.14}$$

The critical points of $V_{(C_1, C_2)}$ occur where $V'_{(C_1, C_2)}(x) = 0$

$$3x^2 + 2C_1 x + C_2 = 0 \tag{10.15}$$

The critical points coalesce where $V''_{(C_1, C_2)}(x) = 0$

$$6x + 2C_1 = 0 \tag{10.16}$$

Now, eliminating x from Eqs. (10.15) and (10.16), we get

$$\frac{1}{3} C_1^2 - \frac{2}{3} C_1^2 + C_2 = 0$$

i.e.,

$$C_2 = \frac{1}{3} C_1^2 \tag{10.17}$$

Equation (10.17) is the equation of a parabola in the $C_1 - C_2$ plane. This is the catastrophe set k for the given V.

When $C_2 < \dfrac{1}{3} C_1^2$, there is one minimum and one maximum for V.

When $C_2 > \dfrac{1}{3} C_1^2$, there are no critical points. As the point $(C_1 - C_2)$ moves along k, the system in equilibrium at the minimum of V would exhibit a catastrophic jump [towards $V = -\infty(x = -\infty)$] as the movement proceeds.

Further, according to Thom's theorem, for systems governed by smooth functions, with at most four parameters but any number of variables, there are essentially only seven possible types of structures for stable catastrophe sets.

The seven catastrophe models available to study catastrophic events are as follows:

Stability Theory

1. The fold
$$V_{C_1}(x) = x^3 + C_1 x$$

2. The cusp
$$V_{(C_1, C_2)}(x) = x^4 + C_1 x^2 + C_2 x$$

3. The swallow tail
$$V_{(C_1, C_2, C_3)}(x) = x^5 + C_1 x^3 + C_2 x^2 + C_3 x$$

4. The elliptic umbilic or hair
$$V_{(C_1, C_2, C_3)}(x_1, x_2) = x_1^3 - 3 x_1 x_2^2 + C_1 (x_1^2 + x_2^2) + C_2 x_1 + C_3 x_2$$

5. The hyperbolic umbilic or breaking wave
$$V_{(C_1, C_2, C_3)}(x_1, x_2) = x_1^3 + x_2^3 + C_1 x_1 x_2 + C_2 x_1 + C_3 x_2$$

6. The butterfly
$$V_{(C_1, C_2, C_3, C_4)}(x) = x^6 + C_1 x^4 + C_2 x^3 + C_3 x^2 + C_4 x$$

7. The parabolic umbilic or the mushroom
$$V_{(C_1, C_2, C_3, C_4)}(x_1, x_2) = x_2^4 + x_1^2 x_2 + C_1 x_1^2 + C_2 x_2^2 + C_3 x_1 + C_4 x_2$$

The catastrophe models with two variables can be visualised using a 3D plot of $V(x_1, x_2)$ as a function of both x_1 and x_2.

EXERCISES

1. For the system
$$\frac{dx}{dt} = x$$
$$\frac{dy}{dt} = -x + 2y$$
sketch the phase portrait and comment on the type of the critical point.

2. For the system $\frac{d^2 x}{dt^2} + 2b \frac{dx}{dt} + a^2 x = 0$, $b \geq 0$ and $a > 0$, describe the nature and stability of the critical point when $b = 0$ and $b = a$.

3. Sketch the phase portrait for Volterra's predator-prey model for $a = b = c = d = 0.5$.

4. Find all the critical points of the following system and sketch its phase portrait:

$$\frac{d^2x}{dt^2} + K\sin x = 0$$

5. Using Liapunov's method, describe the stability of the critical point (0, 0) of the following system:

$$\frac{dx}{dt} = -2xy$$

$$\frac{dy}{dt} = x^2 - y^2$$

6. Test the stability of the trivial solution of the system

$$\frac{dx}{dt} = -y - x^3$$

$$\frac{dy}{dt} = x - y^3$$

using Liapunov's method.

7. For the cusp catastrophe, find the critical points and sketch the catastrophe sets k in the control space.

8. For $V_{(C_1, C_2)}(x_1, x_2) = x_1^3 - 2x_1x_2 + C_1x_1^2 + x_2^2 + C_2x_1$, find the critical points and the nature of the catastrophe set.

9. Explain the applications of catastrophe models.

CHAPTER 11

State Space Models and Transfer Functions

The state space models and transfer functions are simple linear representations of ordinary differential equation models of systems. Such representations are commonly used for designing suitable controllers for control of the actual system under observation. Further, these models are also useful for analysis of the stability of systems at certain operating points. In this chapter, we discuss the procedure for obtaining the state space model of a non-linear system. Further, we discuss the method of converting the state space model to a transfer function model.

11.1 THE STATE-SPACE MODEL

The state space model is a generalised mathematical model of a system, with a set of inputs, outputs and state variables. It is a linear, dynamic, time domain model in matrix vector form. The structure of the state space model (Figure 11.1) is given as

$$\dot{x} = \frac{dx}{dt} = Ax + Bu$$

$$y = Cx + Du$$

where, x is the state variable, u is the input variable, y is the output variable, A is the system matrix, B is the input matrix, C is the output matrix and D is the direct transmission matrix.

Consider a non-linear model with n states, m outputs and k inputs, represented by

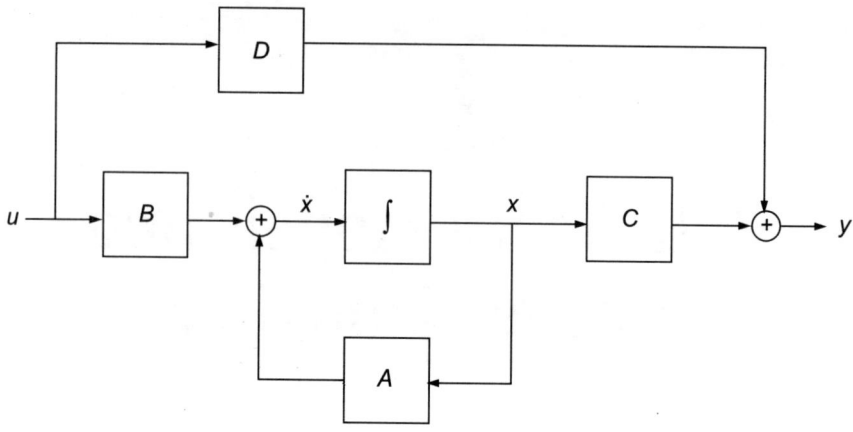

FIGURE 11.1 Block diagram representation of the state space model.

$$\dot{x}_1 = f_1(x_1, x_2, ..., x_n, u_1, u_2, ..., u_k)$$
$$\dot{x}_2 = f_2(x_1, x_2, ..., x_n, u_1, u_2, ..., u_k)$$
$$\vdots$$
$$\dot{x}_n = f_n(x_1, x_2, ..., x_n, u_1, u_2, ..., u_k)$$
$$y_1 = g_1(x_1, x_2, ..., x_n, u_1, u_2, ..., u_k)$$
$$y_2 = g_2(x_1, x_2, ..., x_n, u_1, u_2, ..., u_k)$$
$$\vdots$$
$$y_m = g_m(x_1, x_2, ..., x_n, u_1, u_2, ..., u_k)$$

where, $f_1, f_2, ..., f_n, g_1, g_2, g_m$ are non-linear functions. To obtain the state space representation of this system, the Jacobian matrices A, B, C and D of the state space model are computed as follows:

$$A_{ij} = \left(\frac{\partial f_i}{\partial x_i}\right)_{(\bar{x}, \bar{u})} \text{ is the system matrix.}$$

$$B_{ij} = \left(\frac{\partial f_i}{\partial u_i}\right)_{(\bar{x}, \bar{u})} \text{ is the input matrix.}$$

$$C_{ij} = \left(\frac{\partial g_i}{\partial x_j}\right)_{(\bar{x}, \bar{u})} \text{ is the output matrix.}$$

$$D_{ij} = \left(\frac{\partial g_i}{\partial u_j}\right)_{(\bar{x}, \bar{u})} \text{ is the direct transmission matrix.}$$

where, *ij* represents the *i*th row and *j*th column of the corresponding Jacobian matrices and $\bar{x}, \bar{u}$ are the steady state operating points. Finally, the linear state space model of the non-linear system is represented as

$$\dot{x}' = Ax' + Bu'$$
$$y' = Cx' + Du'$$

where, the deviation variables are defined as the perturbations from their steady state values.

$$x' = x - \bar{x}$$
$$u' = u - \bar{u}$$
$$y' = y - \bar{y}$$

EXAMPLE 11.1 Obtain the state space model of the following linear system, with y as the output variable:

$$\frac{d^3y}{dt^3} + 9 \cdot \frac{d^2y}{dt^2} + 23 \cdot \frac{dy}{dt} + 28y = u(t) \qquad \text{(i)}$$

Solution:
Let
$$x_1 = y$$
$$x_2 = \dot{x}_1 = \frac{dy}{dt}$$
$$x_3 = \dot{x}_2 = \frac{d^2y}{dt^2}$$

Equation (i) can be written as

$$\frac{d^3y}{dt^3} = u(t) - 9\dot{x}_2 - 23\dot{x}_1 - 28x_1$$

Rewriting the equations,

$$\dot{x}_1 = x_2$$
$$\dot{x}_2 = x_3$$
$$\dot{x}_3 = -28x_1 - 23x_2 - 9x_3 + u(t)$$

and
$$y = x_1$$

The state space model of the given system is

$$\begin{pmatrix} \dot{x}_1 \\ \dot{x}_2 \\ \dot{x}_3 \end{pmatrix} = \begin{pmatrix} 0 & 1 & 0 \\ 0 & 0 & 1 \\ -28 & -23 & -9 \end{pmatrix} \begin{pmatrix} x_1 \\ x_2 \\ x_3 \end{pmatrix} + \begin{pmatrix} 0 \\ 0 \\ 1 \end{pmatrix} u(t)$$

$$y = (1 \ 0 \ 0) \begin{pmatrix} x_1 \\ x_2 \\ x_3 \end{pmatrix}$$

The state flow of the given system is shown in the block diagram of the given figure.

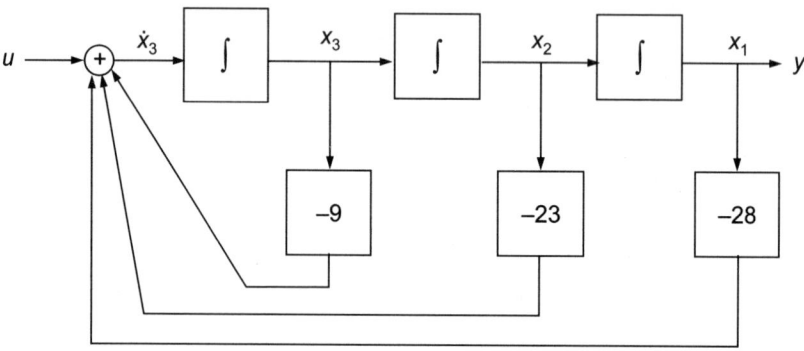

EXAMPLE 11.2 Obtain the linear state space model of the following non-linear model:

$$\frac{ds}{dt} = 0.07 - 0.01 u_1 - 0.07 s - 0.008 u_1 s + k_1$$

$$\frac{du_1}{dt} = -0.099 u_1 + 0.007 u_1 s - 0.001 u_1^2 - k_2$$

This model represents the dynamics of heroin addiction epidemic in a closed geographical area. Here, s is the susceptible population, i.e., the number of people who are not yet addicted to heroin but have the potential to get addicted. u_1 represents the addicted population, i.e., the number of people who are already addicted to the drug. k_1 and k_2 are the control forces for controlling the epidemic. k_1, represents the awareness programmes and k_2 represents the rehabilitation programmes.

Solution: The linear state space model of the system is

$$\begin{pmatrix} \frac{ds'}{dt} \\ \frac{du_1'}{dt} \end{pmatrix} = \begin{pmatrix} -0.07 - 0.008 \bar{u}_1 & -0.01 - 0.008 \bar{s} \\ 0.007 \bar{u}_1 & -0.099 + 0.007 \bar{s} - 0.002 \bar{u}_1 \end{pmatrix} \begin{pmatrix} s' \\ u_1' \end{pmatrix} + \begin{pmatrix} 1 & 0 \\ 0 & -1 \end{pmatrix} \begin{pmatrix} k_1' \\ k_2' \end{pmatrix}$$

$$\begin{pmatrix} y_1 \\ y_2 \end{pmatrix} = \begin{pmatrix} 1 & 0 \\ 0 & 1 \end{pmatrix} \begin{pmatrix} s' \\ u_1' \end{pmatrix}$$

where, $s' = s - \bar{s}$

$$u_1' = u_1 - \bar{u}_1$$
$$k_1' = k_1 - \bar{k}_1$$
$$k_2' = k_2 - \bar{k}_2$$

11.2 CONTROLLABILITY AND REACHABILITY

Controllability

The controllability of a state space model answers the question whether we can steer the state using the control inputs to certain locations in the state space. The issue of controllability concerns whether a given initial state x_0 can be steered to the origin in finite time using the input $u(t)$.

Formally, a state x_0 is said to be controllable if there exists a finite time interval [0, T] and an input $\{u(t), t \in [0, T]\}$ such that $x(T) = 0$. If all the states are controllable, then the system is said to be completely controllable.

Test for controllability: Consider the following state space model:

$$\dot{x} = Ax + Bu$$
$$y = Cx + Du$$

The set of all controllable states is the range space of the controllability matrix $\Gamma_c[A, B]$, where $\Gamma_c[A, B] \triangleq [B \quad AB \quad A^2 \dots A^{n-1}B]$

The model is completely controllable if and only if $\Gamma_c[A, B]$ has full row rank.

Reachability

A state $\bar{x} \neq 0$ is said to be reachable (from the origin) if, given $x(0) = 0$, there exists a finite time interval [0, T] and an input $\{u(t), t \in [0, T]\}$ such that $x(T) = \bar{x}$. If all the states are reachable, the system is said to be completely reachable.

EXAMPLE 11.3 Test the controllability of the following system:

$$\dot{x} = \begin{bmatrix} -3 & 1 \\ -2 & 0 \end{bmatrix} x + \begin{bmatrix} 1 \\ -1 \end{bmatrix} u$$

Solution: The controllability matrix for the given system is

$$\Gamma_c[A, B] = [B, \quad AB] = \begin{bmatrix} 1 & -4 \\ -1 & -2 \end{bmatrix}_{2 \times 2}$$

where, $A = \begin{bmatrix} -3 & 1 \\ -2 & 0 \end{bmatrix}_{2 \times 2}$ and $B = \begin{bmatrix} 1 \\ -1 \end{bmatrix}$

The rank of $\Gamma_c[A, B] = 2 = n$.

Hence, the system is completely controllable.

11.3 OBSERVABILITY

Observability is concerned with the issue of what information can be obtained about the states when one is given measurements of the system output. A system is said to observable if the initial state of the system can be observed or determined using the measured inputs and outputs of the system. The system is said to be completely observable if there exists no non-zero initial state that are unobservable.

Test for observability: Consider the following state space model:

$$\dot{x} = Ax + Bu$$
$$y = Cx + Du$$

The set of all unobservable states is equal to the null space of the observability matrix $\Gamma_0[A, C]$, where

$$\Gamma_0[A, C] \triangleq \begin{bmatrix} C \\ CA \\ \vdots \\ CA^{n-1} \end{bmatrix}$$

The system is completely observable if and only if $\Gamma_0[A, C]$, has full column rank n.

EXAMPLE 11.4 Check whether the following state space model is completely observable:

$$\dot{x} = \begin{bmatrix} -3 & -2 \\ 1 & 0 \end{bmatrix} x + \begin{bmatrix} 1 \\ 0 \end{bmatrix} u$$

$$y = \begin{bmatrix} 1 & -1 \end{bmatrix} x$$

Solution: The observability matrix of the given system is

$$\Gamma_0[A, C] = \begin{bmatrix} C \\ CA \end{bmatrix} = \begin{bmatrix} 1 & -1 \\ -4 & -2 \end{bmatrix}$$

Rank $(\Gamma_0[A, C]) = 2$

Hence, the system is completely observable.

11.4 THE STATE TRANSITION MATRIX

The state transition matrix ϕ is a matrix whose product with the state vector x at an initial time t_0, gives x at a later time t. The state transition matrix can be used to obtain the general solution of linear dynamical systems. The function ϕ is called the state transition matrix because it controls the change of states in the state equation.

Now, let us derive the expression for the state transition matrix of the state space model.

$$\dot{x} = Ax + Bu$$
$$y = Cx + Du$$

Considering a homogeneous case (input free), the state equation is expressed as

$$\dot{x} = Ax(t) \tag{11.1}$$

The solution to this equation is given as

$$x(t) = e^{At} \cdot x(0) \tag{11.2}$$

Taking Laplace transform of Eq. (11.1),

$$sX(s) - x(0) = AX(s)$$
$$(sI - A)X(s) = x(0)$$
$$\Rightarrow \quad X(s) = (sI - A)^{-1}x(0)X(s) \tag{11.3}$$

Taking inverse Laplace transform of Eq. (11.3), we get

$$x(t) = L^{-1}((sI - A)^{-1})x(0) \tag{11.4}$$

From Eq. (11.2) and (11.4), we obtain the expression for the state transition matrix as

$$\phi(t) = e^{At} = L^{-1}[(sI - A)^{-1}] \tag{11.5}$$

Now, let us substitute $t = t_0$ in Eq. (11.2)

$$x(t_0) = e^{At_0} \cdot x(0)$$
$$\Rightarrow \quad x(0) = e^{-At_0} \cdot x(t_0) \tag{11.6}$$

Substituting Eq. (11.6) in Eq. (11.2), we get

$$x(t) = e^{At} \cdot e^{-At_0} \cdot x(t_0)$$
$$= e^{A(t-t_0)} \cdot x(t_0)$$

Since $\phi(t) = e^{A(t)}$,

$$x(t) = \phi(t - t_0) \cdot x(t_0)$$

Now, let us consider the non-homogeneous case (where the inputs are considered). The state space representation is

$$\dot{x} = Ax + Bu \qquad (11.7)$$
$$y = Cx + Du \qquad (11.8)$$

Taking Laplace transform of Eq. (11.7),

$$sX(s) - x(0) = AX(s) + BU(s)$$
$$(sI - A)X(s) = x(0) + BU(s)$$
$$X(s) = (sI - A)^{-1}x(0) + (sI - A)^{-1}BU(s) \qquad (11.9)$$

Taking inverse Laplace transform of Eq. (11.9), we get

$$x(t) = \phi(t)x(0) + \int_0^t \phi(t-\tau) Bu(\tau) d\tau$$

$$\underbrace{x(t) = \phi(t-t_0).x(t_0)}_{\text{Unforced response}} + \underbrace{\int_0^t \phi(t-\tau) Bu(\tau) d\tau}_{\text{Forced response}} \qquad (11.10)$$

Now, substituting Eq. (11.10) in Eq. (11.8), we get the response of the system

$$y(t) = C\phi(t-t_0)x(t_0) + C\int_0^t \phi(t-\tau) Bu(\tau) d\tau + Du(t)$$

11.4.1 PROPERTIES OF STATE TRANSITION MATRIX

The state transition matrix has some fixed properties, which are listed below:
1. $\phi(0) = I$
2. $\phi^{-1}(t) = \phi(-t)$
3. $x(0) = \phi(-t) x(t)$
4. $\phi(t_2 - t_1) \phi(t_1 - t_0) = \phi(t_2 - t_0)$
5. $\phi^k(t) = \phi(kt)$

EXAMPLE 11.5 Compute the state transition matrix of the following system:

$$\begin{pmatrix} \dot{x}_1 \\ \dot{x}_2 \\ \dot{x}_3 \end{pmatrix} = \begin{pmatrix} -1 & 0 & 0 \\ 0 & -2 & 0 \\ 0 & 0 & -3 \end{pmatrix} \begin{pmatrix} x_1 \\ x_2 \\ x_3 \end{pmatrix} + \begin{pmatrix} 1 \\ 1 \\ 1 \end{pmatrix} u$$

$$y = \begin{pmatrix} 6 & -6 & 1 \end{pmatrix} \begin{pmatrix} x_1 \\ x_2 \\ x_3 \end{pmatrix}$$

Solution: The system matrix $A = \begin{pmatrix} -1 & 0 & 0 \\ 0 & -2 & 0 \\ 0 & 0 & -3 \end{pmatrix}$

The state transition matrix is

$$\phi(t) = e^{At} = \begin{pmatrix} e^{-t} & 0 & 0 \\ 0 & e^{-2t} & 0 \\ 0 & 0 & e^{-3t} \end{pmatrix}$$

11.5 VARIOUS STATE SPACE REPRESENTATIONS

As discussed earlier, the state space representation of a linear time-invariant system is given by

$$\dot{x}(t) = Ax(t) + Bu(t)$$
$$y(t) = Cx(t) + Du(t)$$

And for a linear time-variant system, the state space model is given by

$$\dot{x}(t) = A(t)x(t) + B(t)u(t)$$
$$y(t) = C(t)x(t) + D(t)u(t)$$

A non-linear system can be represented using a linear time-variant model as follows:

The synthesis procedure involves finding a description of the non-linear model $\dot{x} = f(x, u)$, to the form

$$\dot{x} = A(\rho)x + B(\rho)u \qquad (11.11)$$

where, ρ is the state-dependent parameter vector.

For example, consider the following non-linear system:

$$\dot{x} = -x^5 + u$$

This system can be converted to the form given by Eq. (11.11) as

$$\dot{x} = -\rho x + u$$

where, $\rho = x^4$.

The state space representation of a linear fractional differential equation can be written as

$$\frac{d^\alpha x}{dt^\alpha} = Ax + Bu \qquad (11.12)$$

172 *Mathematical Modelling of Systems and Analysis*

$$y = Cx + Du \tag{11.13}$$

where, $\alpha = [\alpha_1, \alpha_2, ..., \alpha_n]$, $u \in R^l$ is the input column vector, $x \in R^n$ is the state column vector, $y \in R^p$ is the output column vector, $A \in R^{n \times n}$ is the state matrix, $B \in R^{n \times l}$ is the input matrix, $C \in R^{p \times n}$ is the output matrix and $D \in R^{p \times l}$ is the direct transmission matrix. Equation (11.12) is referred to as the fractional order state equation and Eq. (11.13) is the output equation.

11.6 TRANSFER FUNCTION MODELS

Transfer function models are linear dynamic input-output models in frequency domain. The transfer function of a linear time-invariant system is expressed as the ratio of two polynomials. For a single input single output system, the transfer function $G(s)$ is of the form

$$G(s) = \frac{b_0 s^n + b_1 s^{n-1} + b_2 s^{n-2} + ... + b_{n-1} s + b_n}{a_0 s^m + a_1 s^{m-1} + a_2 s^{m-2} + ... + a_{m-1} s + a_m}$$

where, $[b_0, b_1, ..., b_n; a_0, a_1, ..., a_m]$ are polynomial coefficients, and ω is the angular frequency. The transfer function models of systems provide a compact representation of the system and are suitable for frequency-domain analysis, stability analysis and for the design of suitable controllers for controlling the actual system. However, the transfer functions suffer from neglection of initial conditions. Further, the transfer functions can be derived from the state space models as follows:

$$\dot{x}(t) = Ax(t) + Bu(t) \tag{11.14}$$

$$y(t) = Cx(t) + Du(t) \tag{11.15}$$

Taking Laplace transform of Eq. (11.14) yields

$$sX(s) - x(0) = AX(s) + BU(s)$$

$$(sI - A)X(s) = x(0) + BU(s)$$

$$X(s) = (sI - A)^{-1} x(0) + (sI - A)^{-1} BU(s) \tag{11.16}$$

Substituting Eq. (11.16) in the output Eq. (11.15), we get

$$Y(s) = CX(s) + DU(s)$$

$$Y(s) = C[(sI - A)^{-1} BU(s)] + DU(s) \tag{11.17}$$

The transfer function is defined as the ratio of the output to the input of the system. Hence, the transfer function is represented as

$$G(s) = \frac{Y(s)}{U(s)} \tag{11.18}$$

Substituting Eq. (11.17) in Eq. (11.18), we get

$$G(s) = C(sI - A)^{-1}B + D$$

where, $s = j\omega$.

11.6.1 Transfer Function Models of Fractional Order Systems

Recalling the linear time invariant fractional state space model given by

$$D^\alpha x = Ax + Bu \quad (11.19)$$
$$y = Cx + Du \quad (11.20)$$
$$0 < \alpha < 1$$

Taking Laplace transform of Eq. (11.19), we get

$$s^\alpha X(s) - s^{\alpha-1}x(0) = AX(s) + BU(s)$$

$$X(s) = (s^\alpha I - A)^{-1}BU(s) + (s^\alpha I - A)^{-1}s^{\alpha-1}x(0) \quad (11.21)$$

Now, assume that the initial conditions are zero, i.e., $x(0) = 0$. Then, Eq. (11.21) becomes

$$X(s) = (s^\alpha I - A)^{-1}BU(s) \quad (11.22)$$

Substituting Eq. (11.22) in Eq. (11.20), we get

$$Y(s) = G(s) \, U(s)$$

where, $G(s) = C(s^\alpha I - A)^{-1}B + D$

11.6.2 Advantages of State Space Model over Transfer Function Models

The state-space representation serves as an alternative to transfer functions, since it can easily handle multi input multi output (MIMO) systems and is not limited to time invariant systems. Further, the state space representation gives a better insight into the inner dynamics of the system. Also, the system properties such as controllability and observability can easily be defined and determined using the state space model. Unlike the transfer function models, in state space models, the initial conditions are not neglected, and is best suited for optimisation of system performance.

11.7 STABILITY ANALYSIS USING STATE SPACE MODELS AND TRANSFER FUNCTIONS

Internal Stability

Internal stability deals with the boundedness and asymptotic behaviour (as $t \to \infty$) of the solutions of the system.

Consider the following system:
$$\dot{x} = Ax$$
Its solution is $x(t) = e^{At}x(0)$.
The system is stable if
$$\|e^{AT}\| \leq \gamma, \ \forall t \geq 0, \ \gamma > 0$$
and asymptotically (or exponentially) stable if
$$\|e^{AT}\| \leq \gamma e^{-\lambda t}, \ \forall t \geq 0, \ \gamma > 0, \ \lambda > 0$$
In other words, the system $\dot{x} = Ax$ is stable if and only if
$$\text{Real}(\lambda_i) \leq 0, \ i = 1, 2, \ldots, n$$
where, λ_i denotes the eigenvalues of the system matrix A.
And the system is asymptotically stable if
$$\text{Real}(\lambda_i) < 0, \ i = 1, 2, \ldots, n$$
For example, consider the system given by
$$\dot{x} = Ax$$
where, $A = \begin{bmatrix} -1 & 0 & 3 & 1 \\ 0 & -2 & 1 & 0 \\ 0 & 0 & -3 & 1 \\ 0 & 0 & 0 & -3 \end{bmatrix}$

The eigenvalues of A are $-1, -2, -3$ and -3. Since all eigenvalues of A are negative, the system is asymptotically stable.

Consider another example where the system matrix is
$$A = \begin{bmatrix} -1 & 0 & 3 & 1 \\ 0 & 1 & 1 & 0 \\ 0 & 0 & 0 & 0 \\ 0 & 0 & 0 & 0 \end{bmatrix}$$

The eigenvalues of A are $-1, 1, 0$ and 0. Since there is a positive eigenvalue, the system is unstable.

Bounded input bounded output stability (BIBO) stability

The linear system given by the transfer function
$$Y(s) = H(s)\,U(s) \tag{11.23}$$
is BIBO stable if for every bounded input $u(t)$, the output $y(t)$ is bounded equivalently. For every $K_u > 0$, there is $K_y > 0$ such that

$$\|u(t)\| \le K_u, \forall t \ge 0 \implies \|y(t)\| \le K_y, \forall t \ge 0$$

Taking inverse Laplace transform of Eq. (11.23), we get

$$y(t) = \int_0^t H(t-\tau)u(\tau)d\tau$$

The system is BIBO stable if and only if $\int_0^\infty \|H(t)\| dt < \infty$.

When $H(s) = C(sI - A)^{-1}B + D$, the poles of $H_{ij}(s)$ are roots of $|sI - A| = 0$, which are the eigenvalues of A. Therefore, $H(s)$ is BIBO stable if and only if all poles of $H(s)$ have negative real parts.

Relationship between asymptotic stability and BIBO stability

1. $\dot{x} = Ax$ is asymptotically stable if all eigenvalues of A have negative real parts.
2. $H(s) = C(sI - A)^{-1}B + D$ is BIBO stable if all poles of all elements of $H(s)$ have negative real parts. The poles of $H(s)$ are eigenvalues of A.

Asymptotic stability implies BIBO stability, however, BIBO stability does not imply asymptotic stability. Some eigenvalues of A may not appear as parts of $H(s)$. If such eigenvalues have non-negative real parts, then the system is BIBO stable but not asymptotically stable.

Stability of fractional order systems

Consider the fractional order transfer function $G(s) = C(s^\alpha I - A)^{-1}B + D$. The elements of $G(s)$ are transfer functions with a common polynomial denomination given by

$$\phi(s^\alpha) = \det(s^\alpha I - A)$$

Let P_i be the poles of the system $G(s)$. The poles are defined as the solution of the equation $\phi(s^\alpha) = 0$, which is obtained as

$$P_i = \lambda_i^{1/\alpha}$$

where λ_i, $1 \le i \le n$ are the eigenvalues of matrix A. The stability condition in the BIBO sense is attained if the poles of the system lie in the negative half plane of the s-complex plane $\left(|\arg(P_i)| > \dfrac{\pi}{2}\right)$.

EXERCISES

1. Find the state transition matrix of the following state space model:

$$\dot{x} = \begin{pmatrix} -5 & 0 & 0 \\ 0 & -7 & 0 \\ 0 & 0 & -9 \end{pmatrix} x + \begin{pmatrix} 1 \\ 1 \\ 1 \end{pmatrix} u$$

$$y = (2 \ -2 \ 1)x$$

2. Mention the properties of state transition matrix.
3. State the relation between BIBO stability and asymptotic stability.
4. Convert the following state space model into transfer function model:

$$D^{0.2}x = \begin{pmatrix} -3.678 & 0 \\ 0.768 & -3.531 \end{pmatrix} x + \begin{pmatrix} 7 & 0.698 \\ -2.787 & 0 \end{pmatrix} u$$

$$y = (0 \ 1)x$$

5. Define controllability and observability of a system.
6. Comment on the stability of the system with system matrix $A = \begin{pmatrix} -1.8124 & -0.2324 \\ 9.6837 & 1.4697 \end{pmatrix}$.

7. Derive the continuous state space model of a mass-spring-dashpot system, as shown in the figure, with $f(t)$ as the applied force and $y(t)$ as the displacement.

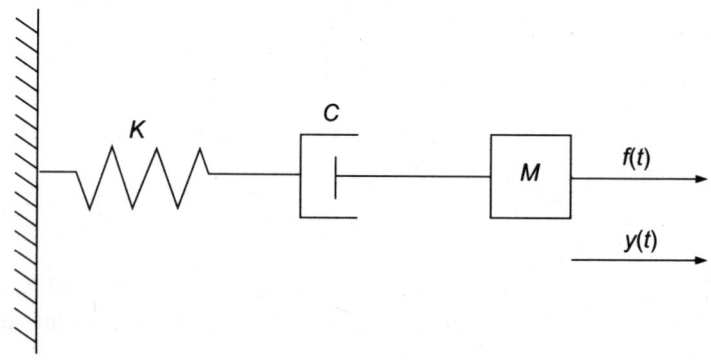

8. Derive the linear state space model of the following RLC network:

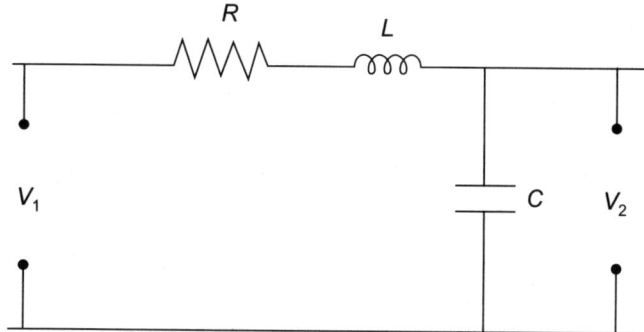

where, V_1 is the input variable and V_2 is the output variable.

9. Convert the following state space model into a transfer function model:

$$A = \begin{pmatrix} -2.4048 & 0 \\ 0.8333 & -2.2381 \end{pmatrix}, \quad B = \begin{pmatrix} 7 & 0.5714 \\ -1.117 & 0 \end{pmatrix}, \quad C = (0 \quad 1), \quad D = 0$$

10. Comment on the stability of the following system:

$$\dot{x} = \begin{pmatrix} 1 & 2 & 1 \\ 2 & 1 & 1 \\ 3 & 4 & 5 \end{pmatrix} x + [0 \ 0 \ 1] u$$

$$y = (5 \ 1 \ 0)x + 0.u$$

11. For the following bio-reactor model, derive the linear state space equations and the transfer function model. (X is the output variable; S_{in} is the input variable.)

$$\frac{dX}{dt} = \mu X - DX$$

$$\frac{dS}{dt} = -\frac{\mu}{Y}X + D(S_{in} - S)$$

where, X is cell concentration, S is substrate concentration, μ is specific growth rate, D is dilution rate, Y is yield coefficient and S_{in} is input substrate concentration.

CHAPTER 12
System Identification

System identification is the art and science of developing mathematical models of dynamic systems from measured input-output data. It involves developing the black box and grey box models of systems. In some systems, the internal dynamics, components and relationships are not visible, and hence, the models based on first principles (white box) cannot be constructed. Modelling of such systems involves experimenting with the system by exciting the system using known inputs and collecting its outputs. Now, the task of developing a model involves finding the mathematical relationship between the inputs and the outputs.

Before collecting the input-output data, it is important to analyse the requirement or need of the model to be developed, and the selection of model types and structures should be based on the requirements. The overall flow chart of the identification process is shown in Figure 12.1.

Further, the identification process can be online in the case of time-varying systems, where the model parameters are updated at each sampling instant, and it can be offline in the case of time-invariant systems, where the experimental data is fully collected, and finally, the model parameters are estimated. The block diagrams representing the process of online and offline identification are shown in Figures 12.2 and 12.3, respectively.

12.1 EXPERIMENTAL DESIGN

The first step of system identification involves collection of input-output data from the real-world system. Here comes the problem of experimental design or the input design problem. According to the *principle of optimal design of experiments*, we seek to design experiments which give the maximum possible information about the system.

In order to develop an accurate model from the measurement or observation data, it is important to design a suitable input which can excite the system so that maximum information about the system appears in the output of the system.

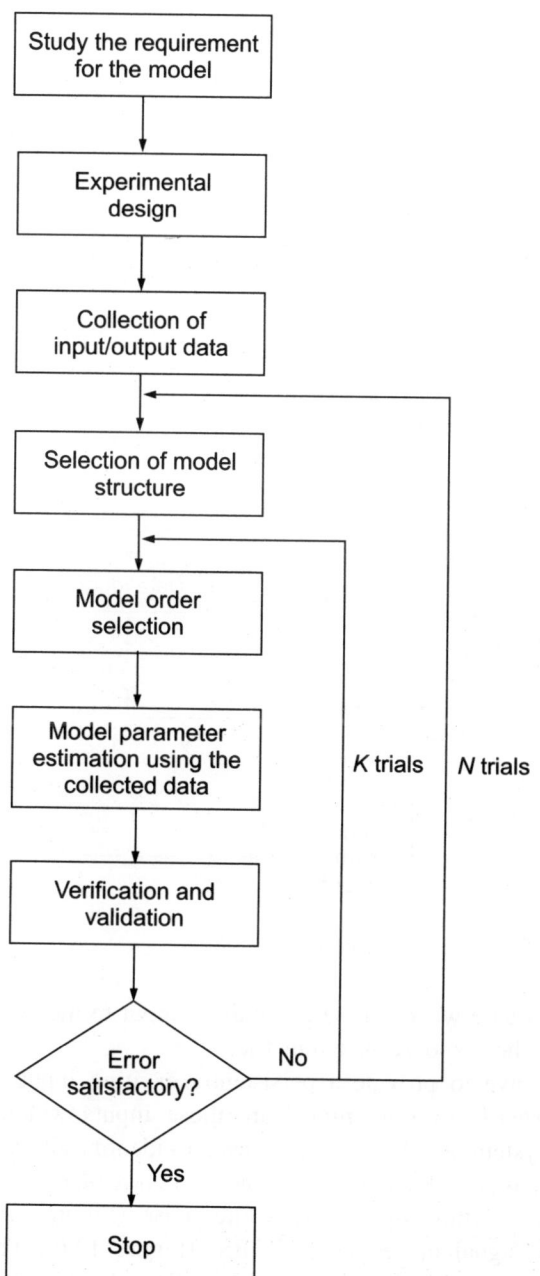

FIGURE 12.1 Flow chart of system identification process.

Before exciting the system, we need to first find out whether the system is linear or not. This can be done by analysing the system responses for various input signal amplitudes and checking whether superposition theorem holds

180 Mathematical Modelling of Systems and Analysis

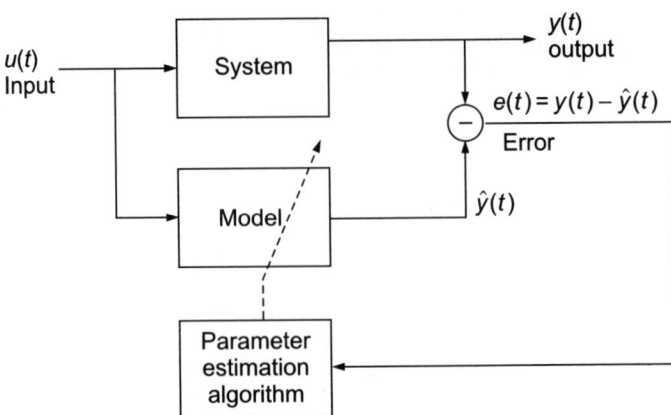

FIGURE 12.2 Online identification process.

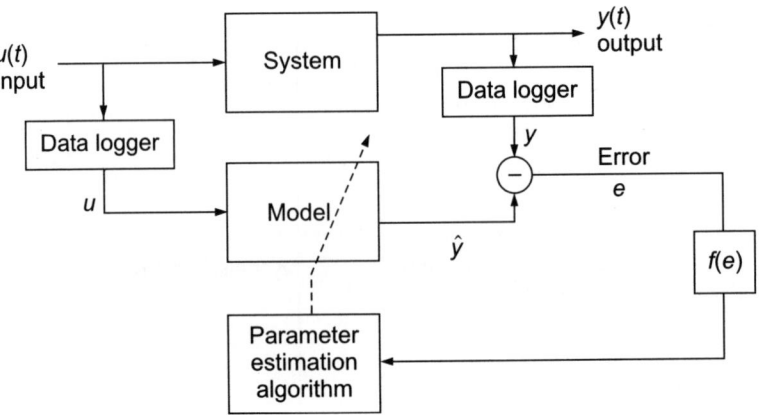

FIGURE 12.3 Offline identification process.

good. Also, a square wave can be given as an input to the system and we can check whether the response is symmetric.

Now, we have to provide a persistently exciting input sequence to the system. *Persistently exciting inputs* are those inputs which can excite all modes of the system, and hence, can be used to identify all parts of the model. Such a signal has rich dynamics and is not too correlated. Common examples of persistently exciting signals are white noise (Figure 12.5) and pseudo random binary signal or sequence (PRBS) (Figure 12.6). These signals are low correlation signals and the output data collected using these input signals provides unique information about all the model parameters, in contrast to the conventional step signal (Figure 12.4), which is a highly correlated signal and provides unique information only about the process gain.

System Identification **181**

FIGURE 12.4 A step signal.

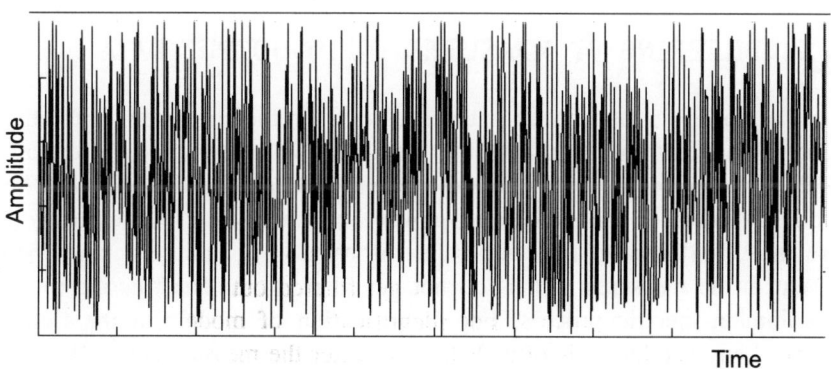

FIGURE 12.5 White noise.

A PRBS is an approximation to the white noise, which is a Gaussian noise, uncorrelated, with constant variance and zero mean. The PRBS is a means of approximating the white noise using only two levels (high/low). A PRBS can be generated using a set of shift registers.

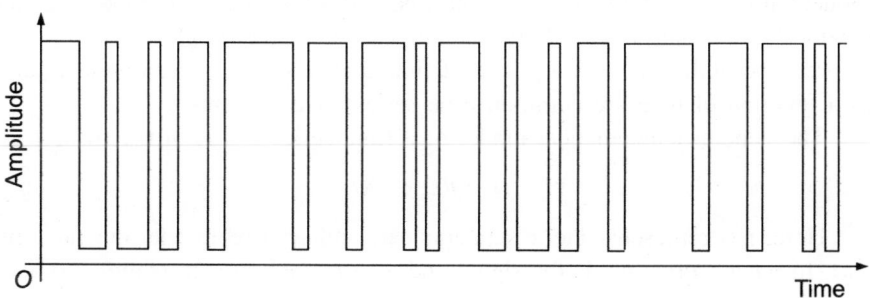

FIGURE 12.6 A PRBS sequence.

In the next part of the experimental design problem comes the question of how much data must be collected for accurate identification of the model parameters. The accuracy of the identification usually improves as the number of data points increases. Hence, the test duration must be longer to ensure good identification. One can fix the test duration to be at least as long as the settling time of the system.

However, it is not possible to excite the system using a PRBS or white noise signal for long test durations for all systems. For example, consider identification of physiological and biological systems. It is not always possible to excite such systems using persistently exciting signals. Hence, the identification of these systems involves conducting experiments of short test durations using very high frequency input signals. The idea behind exciting physiological systems (such as organs of the human body) using high frequency signals is that we can complete the experimentation and data collection before the system gets affected by the input signal itself.

12.2 FILTERING OF MEASURED INPUT-OUTPUT DATA

In some cases, it may be advantageous to filter the measured input/output signals, since filtering can help the identification process by concentrating on the frequency range of interest and can also reduce the effect of high frequency noise or low frequency power line interference noise. However, care must be taken so that important information about the actual system present in the measured signals do not get filtered out. The filtering process is a domain-specific process. For identification of models in the field of electrical engineering, it is often helpful to filter the measurements. However, in the identification of biomedical systems, filtering may result in the loss of important physiological information.

12.3 TIME SERIES

Consider a time series $x(t)$ which is collected by making a series of observations sequentially in time t. Such data is collected and is highly useful in the fields of meteorology, geophysics, biology, economics, commerce, etc. For example, the record of rainfall data over a period of time and records of ocean temperature as a function of time are common time series measurements.

The general mathematical model of a time series $x(t)$ is of the form

$$x(t) = f(t) + w(t)$$

Here, $f(t)$ represents the systematic part and $w(t)$ represents the random part. In other words, $f(t)$ is the signal and $w(t)$ is the noise. In reality, $f(t)$ and $w(t)$ are not separately measurable.

A model in which the effect of time is represented using a random variable is known as a *stochastic process*. The systematic part is represented by a non-random or deterministic function of time. The different types of variations in the systematic part can be described by the following components:

Trend: The trend is a slowly varying function of time. For example, the exponential growth of a population is a trend. The trend is usually represented by means of a lower order polynomial.

Cyclic component: It is a long time movement which is regular and is caused by fluctuations. It is quasi periodic in nature.

Seasonal component: The seasonal component is composed of regular variations caused by rhythmic activities such as changing seasons. Both the cyclic and seasonal components can be represented using periodic functions. Trigonometric functions are generally used for modelling seasonal components.

12.4 SELECTION OF MODELS

Time series models

1. Purely random process (white noise process): A completely random process $\{y_t\}$ has

$$E\{y_t\} = m$$

where, m may be assumed to be 0, and

$$E\{y_t \; y_{t+k}\} = \begin{cases} \sigma^2, & k = 0 \\ 0, & k \neq 0 \end{cases}$$

2. First order Markov process: The structure of the first order Markov process is represented by a linear stochastic difference equation as follows:

$$y_t + \alpha_1 y_{t-1} = e_t, \quad |\alpha_1| < 1$$

where $\{e_t\}$ is purely random which may have $m = 0$ and $\sigma = 1$.

3. Moving average (MA) process: The MA process is represented as:

$$y_t = a_0 e_t + a_1 e_{t-1} + \ldots + a_i e_{t-i}$$

where, $[a_0 \; a_1 \; \ldots \; a_i]$ are real constants and $\{e_i\}$ is a purely random process with mean 0 and variance σ^2.

4. The Yule process: The Yule process is given by

$$y_t + a_1 y_{t-1} + a_2 y_{t-2} = e_t$$

Linear Models

Before discussing some of the common time series models used for system identification, let us define a general model structure for linear time series and black box models. The general structure of linear single input single output commonly used for black box identification is given by

$$A(q)y(t) = \frac{B(q)}{F(q)} u(t - n_k) + \frac{C(q)}{D(q)} e(t)$$

where, $u(t)$ and $y(t)$ are the inputs and outputs, respectively, and e is generally a random noise sequence. A, B, C, D and F are polynomials represented in terms of the backward shift operator q.

$$A(q) = 1 + a_1 q^{-1} + \ldots + a_{na} q^{-na}$$
$$B(q) = b_1 + b_2 q^{-1} + \ldots + b_{nb} q^{-nb+1}$$
$$C(q) = 1 + c_1 q^{-1} + \ldots + c_{nc} q^{-nc}$$
$$D(q) = 1 + d_1 q^{-1} + \ldots + d_{nd} q^{-nd}$$
$$F(q) = 1 + f_1 q^{-1} + \ldots + f_{nf} q^{-nf}$$

1. Autoregressive (AR) model: The AR model is a time series model with no exogenous input $u(t)$. The structure of the AR model is given by

$$A(q)\, y(t) = e(t)$$
$$(1 + a_1 q^{-1} + \ldots + a_{na} q^{-na})\, y(t) = e(t)$$
$$y(t) + a_1 y(t-1) + \ldots + a_{na} y(t-na) = e(t)$$
$$y(t) = e(t) - a_1 y(t-1) - a_2 y(t-2) - \ldots - a_{na} y(t-na)$$

The AR model specifies the value of y at the current time step based on the values of y at the previous time steps and the identification of the AR model involves estimation of the coefficients of polynomial $A(q)$ using the measured sequence $y(t) = [y(1)\ y(2)\ \ldots\ y(k)]$.

2. Autoregressive moving average (ARMA) model: The ARMA model is a strictly proper discrete transfer function and the structure of the model is defined as

$$A(q)\, y(t) = C(q)\, e(t)$$

or

$$\frac{y(t)}{e(t)} = \frac{C(q)}{A(q)}$$

where, the numerator term is the 'moving average' of error or noise and the denominator term is the 'autoregression' of the output.

3. **Autoregressive exogenous input (ARX) model:** The ARX model is an input-output model (Figure 12.7), and the model structure is given by

$$A(q)y(t) = B(q)u(t-i) + e(t)$$

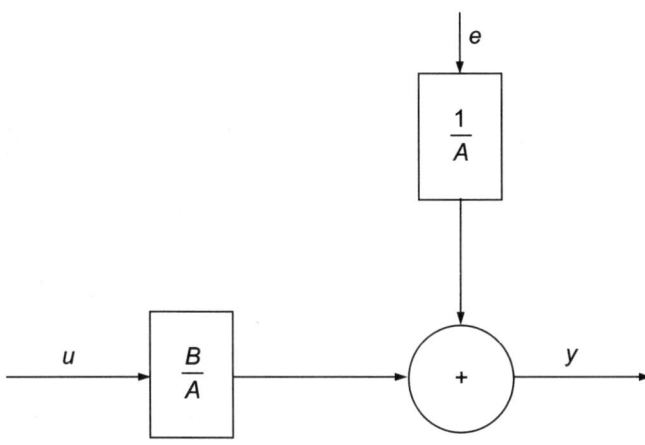

FIGURE 12.7 Block diagram of ARX model.

4. **Autoregressive moving average exogenous input (ARMAX) model:** The structure of ARMAX model is given by

$$A(q)y(t) = B(q)u(t-i) + C(q)e(t)$$

The block diagram of ARMAX model is given in Figure 12.8.

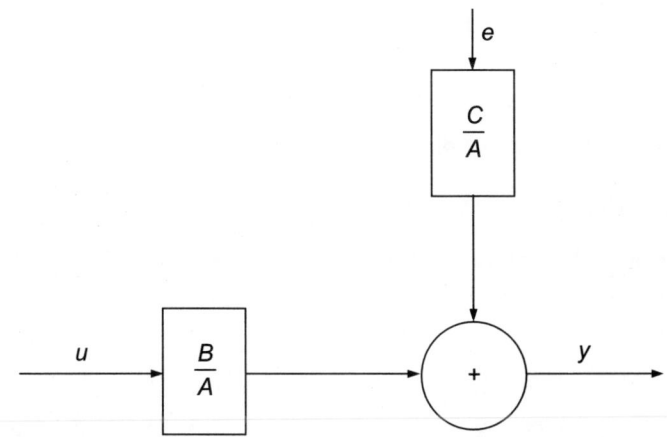

FIGURE 12.8 Block diagram of ARMAX model.

5. **Output error (OE) model:** The OE model is represented by

$$y(t) = \frac{B(q)}{F(q)} u(t-i) + e(t)$$

Figure 12.9 shows the structure of OE model.

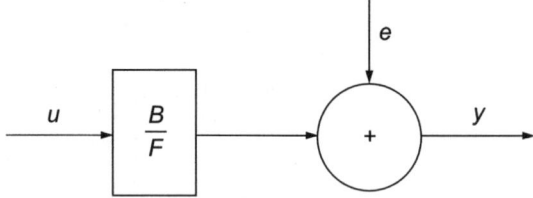

FIGURE 12.9 Structure of OE model.

6. Box-Jenkins (BJ) model: The structure of BJ model (see Figure 12.10) is defined as

$$y(t) = \frac{B(q)}{F(q)} u(t-i) + \frac{C(q)}{D(q)} e(t)$$

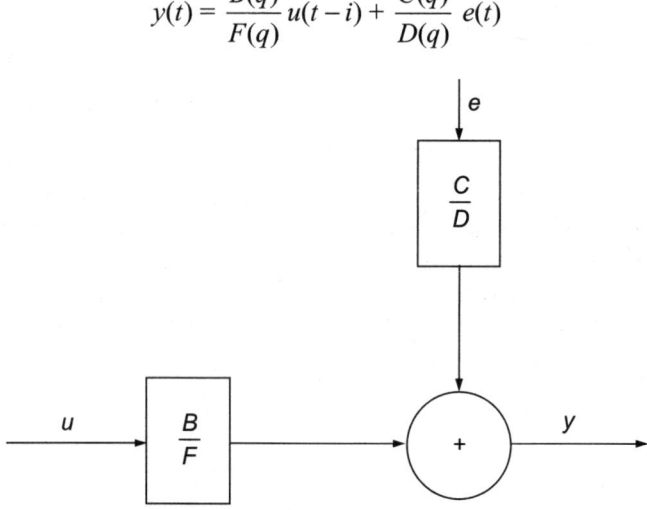

FIGURE 12.10 Structure of BJ model.

Non-linear Models

1. Non-linear ARX model: The non-linear ARX model consists of three blocks, namely, a regressor block, a non-linear function block and a linear function block, as shown in Figure 12.11.

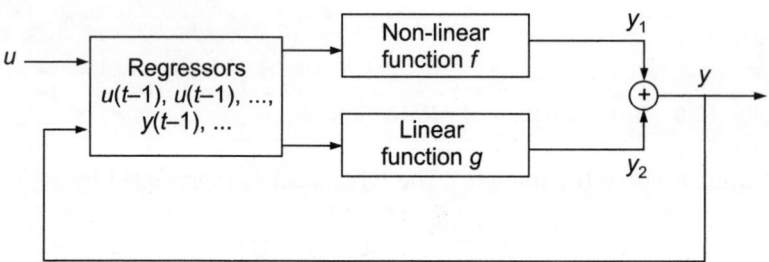

FIGURE 12.11 The non-linear ARX model.

Let u be the input and y be the output of the non-linear ARX model. The regressors block computes the regressors from the current and past input values and past output values. Now, the regressors of the output u and output y act as inputs to the non-linear and linear function blocks.

The output of the non-linear function block is given by

$$y_1 = f(u(t), u(t-1), y(t-1), \ldots, \theta)$$

where f is a function and θ is the set of parameters of the non-linear function.

The output of the linear function block is given as:

$$y_2 = g(u(t), u(t-1), y(t-1), \ldots, \psi)$$

where, g is a linear function and ψ is the set of parameters of g.

The total output of the non-linear ARX model is given by

$$y(t) = y_1(t) + y_2(t)$$

The identification of the non-linear ARX model involves estimation of the model parameters θ and ψ from the available set of measurements of u and y.

2. Hammerstein model: The Hammerstein model is a block-oriented input-output model, with two different blocks, namely, the input non-linearity block and the linear model block, as shown in Figure 12.12.

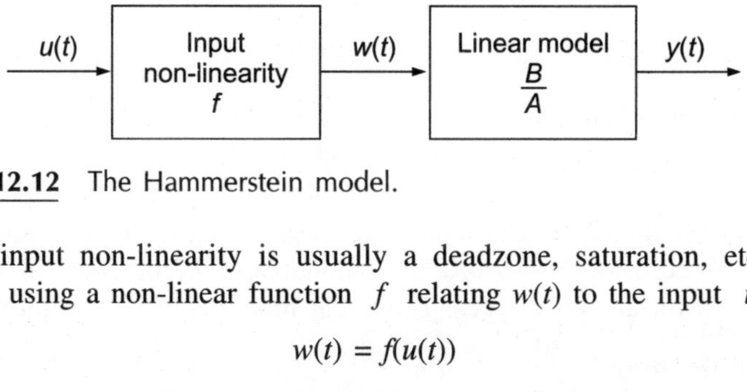

FIGURE 12.12 The Hammerstein model.

The input non-linearity is usually a deadzone, saturation, etc. and is modelled using a non-linear function f relating $w(t)$ to the input $u(t)$.

$$w(t) = f(u(t)) \tag{12.1}$$

The relation between the output $y(t)$ and the intermediate value $w(t)$ is modelled using a linear transfer function $\dfrac{B(q)}{A(q)}$.

$$y(t) = \frac{B(q)}{A(q)} w(t) \tag{12.2}$$

Substituting Eq. (12.1) in Eq. (12.2), we get

$$y(t) = \frac{B(q)}{A(q)} f(u(t)) \tag{12.3}$$

Now, even though the intermediate values $w(t)$ are not measured, the identification of the Hammerstein model can be performed using Eq. (12.3), with measurements of the outputs $y(t)$ and inputs $u(t)$.

3. Weiner model: Similar to the Hammerstein model, the Weiner model consists of two blocks, namely, a linear model block and an output non-linearity block, as shown in Figure 12.13.

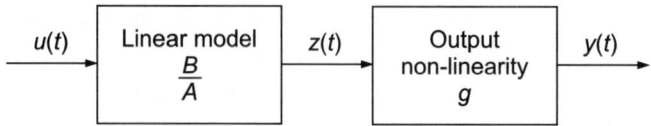

FIGURE 12.13 The Weiner model.

In most of the real-world systems, the output non-linearity is usually a dead time or transportation delay type of non-linearity.

The linear model is a linear transfer function B/A and the output of the linear model is given by

$$z(t) = \frac{B(q)}{A(q)} u(t) \qquad (12.4)$$

Now, $z(t)$ is given as the input to the output nonlinearity block, which produces the final output $y(t)$, which is given by

$$y(t) = g(z(t)) \qquad (12.5)$$

4. Hammerstein–Weiner model: The Hammerstein–Weiner model is a continuation of the Hammerstein model and the Weiner model, and here, both the input and output non-linearity are considered along with the linear model, as shown in Figure 12.14. The input non-linearity and the output non-linearity can be modelled using higher order polynomials, artificial neural networks or any other non-linear model types. The linear model is usually a linear transfer function of the form $\frac{B(q)}{A(q)}$.

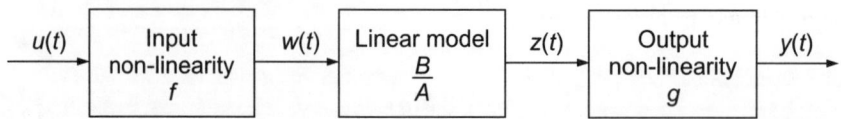

FIGURE 12.14 The Hammerstein–Weiner model.

The output of the input non-linearity block is given by

$$w(t) = f(u(t)) \qquad (12.6)$$

The output of the linear model block is given by

$$z(t) = \frac{B(q)}{A(q)} w(t) \qquad (12.7)$$

The output of the non-linearity block is given by

$$y(t) = g(z(t)) \qquad (12.8)$$

where, f and g are non-linear functions. The identification of Hammerstein–Weiner models from the measured input and output values involves estimation of the parameters of the functions f and g and the coefficients of the polynomials $A(q)$ and $B(q)$.

5. Artificial neural network: Artificial neural networks (ANN) is a branch of artificial intelligence, which has been of wide interest for researchers from various domains and finds applications in almost all fields of science and engineering. ANNs are models that mimic the structure of the human brain (Hinton, 1992). They contain a series of mathematical equations that are used to simulate biological processes such as learning and memory. The developer of a neural network model needs to identify, prior to training, which variables might potentially be important predictors of an outcome. Developers need only a basic understanding of the structure of the neural network and the parameters known as the weights and bias, which can be adjusted for developing the required model (White, 1989; Astion and Wilding, 1992; Tu, 1996).

ANNs can be used for the development of non-linear models of various systems, and in medical and biological modelling problems, ANNs provide a new alternative to logistic regression, which is the most commonly used method for developing predictive models in medicine. Neural networks are able to identify complex non-linear relationships between dependent and independent variables, and have the ability to detect all possible interactions between the associated variables. Further, several training algorithms are available for the development of ANN models (Tu, 1996). Since neural networks are not constrained by a predefined mathematical relationship between dependent and independent variables, they have the ability to model any arbitrarily complex non-linear relationship (White, 1989; Tu, 1996).

A neural network consists of interconnected layered structure comprising a flexible network of weights and transfer functions. Data enters at the input and passes through the network, layer by layer, until it arrives at the output. The parameters of a network are adjusted by training the network on a set of reference data called *training set*. For a particular set of training data consisting of the measured inputs and outputs of a real-world system, the weights of the neural network can be estimated so that the output of the network resembles the real-world outputs. This process of optimising the weights through the minimization of differences between the measured outputs and the network output is known as the *training* of the networks. Once when the weights of the neural networks have been determined, the ANN model can be used for predicting the outputs using new input variables, which are not present in the training set. The weight matrix of the network is analogous with the storage of information in the brain.

Training of the network can be performed using back propagation training algorithm. Figure 12.15 illustrates the architecture of a feedforward neural network consisting of one hidden layer, with $x_1, x_2, ..., x_n$ as inputs and $\hat{y}$ as the output.

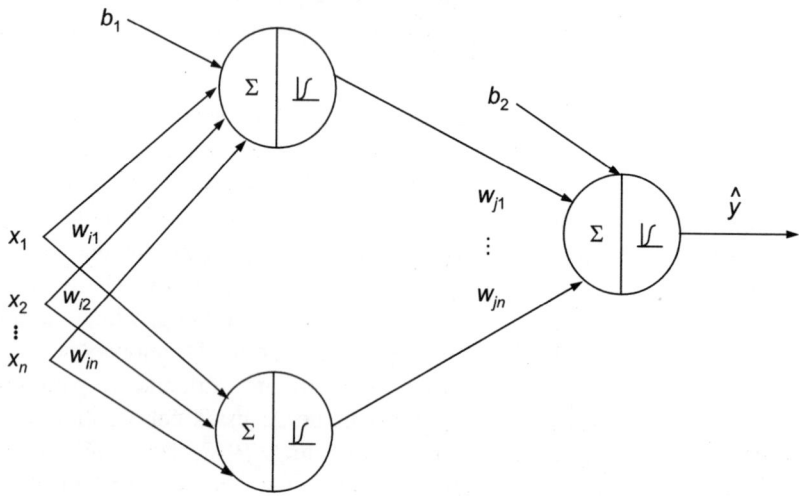

FIGURE 12.15 Architecture of feedforward neural network.

Mathematically, the functionality of a hidden neuron is described by:

$$z_j = \sigma\left(\sum_{j=1}^{n} w_j x_j + b_j\right)$$

where, the weights $\{w_j, b_j\}$ are symbolised with the arrows moving into the neuron. Mathematically, the output of the neural network is given by

$$\hat{y}(\theta) = \sum_{i=1}^{nh} w_i \sigma\left(\sum_{j=1}^{n} w_{ij} x_j + b_{ji}\right) + b_j$$

where, n is the number of inputs and nh is the number of neurons in the hidden layer, $\theta = \{w_{ij}, b_{ji}, w_j, b_j\}$ are the parameters of the ANN model and $\sigma(x)$ is the non-linear activation function. While training the ANN, its parameters are adjusted incrementally until the training errors are minimal.

The non-linear activation function of the network is usually a smooth non-linear function such as a tan sigmoid function given by

$$\sigma(x) = \frac{1}{1+e^{-x}}$$

The network can be trained using the famous back-propagation learning algorithm. The performance of the trained network is usually evaluated using correlation and error analysis. Usually, for a given problem, a number of ANN models are developed and the network configuration with the best performance is finally selected as the model of the system under investigation.

Further, the flow chart of back-propagation algorithm for training the neural networks is presented in Figure 12.16.

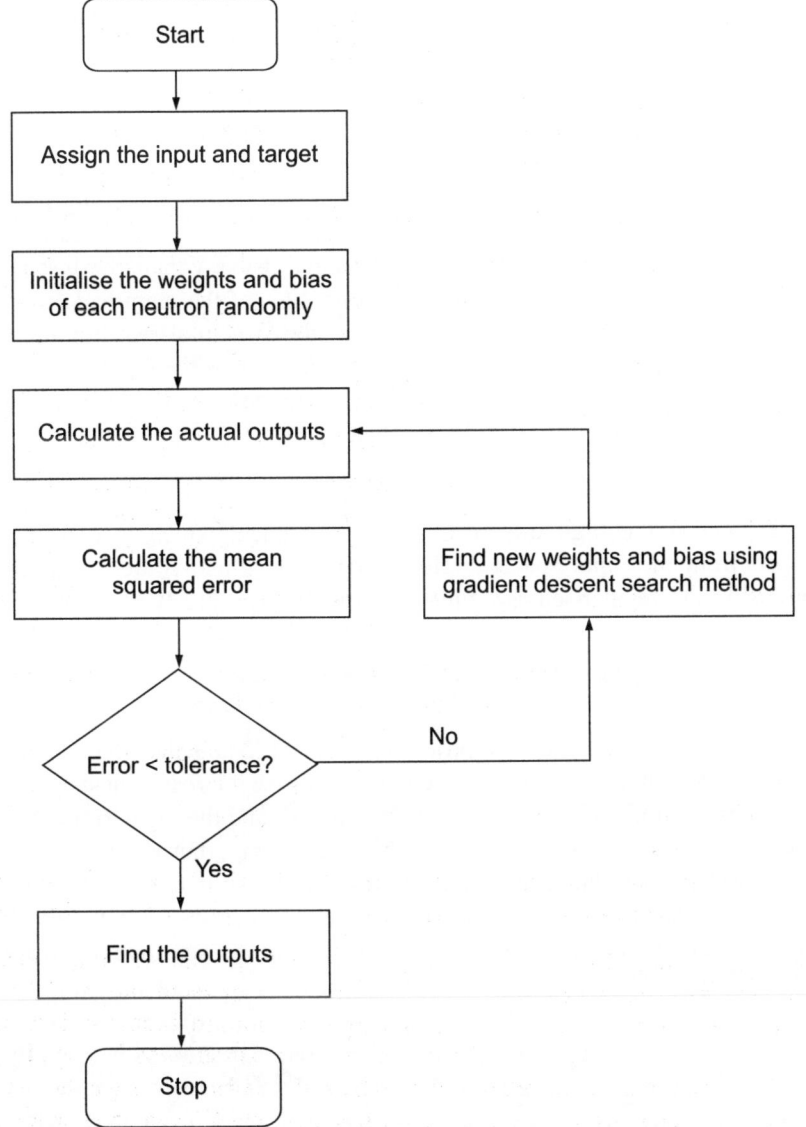

FIGURE 12.16 Flow chart of back-propagation algorithm for training the neural network.

6. Support vector machines (SVMs): Support vector machines are special type of learning algorithms characterised by the capacity control of the decision function. Established on the unique theory of the structural risk minimisation principle to estimate a function by minimising an upper bound of the generalisation error, SVMs are shown to be very resistant to the overfitting problem, eventually achieving high generalisation performance in solving various time series modelling problems (Vapnik, 1996). Support vector machines can be utilized for both classification and regression problems. A SVM used for linear or nonlinear regression is known as a Support Vector Regression (SVR) model.

Formulation of SVR model: Consider a time series X_k and the lag time series from X_k is formed as X_{k-1}. Further, the SVR model can be developed using X_{k-1} as input vectors and X_k as the target vectors.

SVR estimates the regression using a set of linear functions that are defined in a high dimensional space. The regression estimation is carried out by risk minimisation, where the risk is measured using Vapnik's ε-insensitive loss function. SVMs use a risk function consisting of the empirical error and a regularisation term, which is derived from the structural risk minimisation principle. Given a set of data points $G = \{(x_i; d_i)\}_i^n$ (x_i is the input vector, d_i is the desired value and n is the total number of data points), SVMs approximate the function using Eq. (12.9) (Tay and Cao, 2001):

$$y = f(x) = w\phi(x) + b \qquad (12.9)$$

where, $\phi(x)$ is the high dimensional feature space, which is non-linearly mapped from the input space x. The coefficients w and b are estimated by minimising the regularised risk function given by Eq. (12.10).

$$R_{\text{SVMs}}(C) = C\frac{1}{n}\sum_{i=1}^{n} L_g(d_i, y_i) + \frac{1}{2}\|w\|^2 \qquad (12.10)$$

where, the first term is the empirical error or risk and the second term is the regularisation term. C is referred to as the regularised constant and it determines the trade-off between the empirical risk and the regularisation term. Increasing the value of C will result in the relative importance of the empirical risk with respect to the regularisation term to grow. ε is called the tube size and it is equivalent to the approximation accuracy placed on the training data points. The empirical error or risk is measured by the ε-insensitive loss function given by Eq. (12.11).

$$L_g(d, y) = \begin{cases} |d-y|-\varepsilon & |d-y| \geq \varepsilon \\ 0 & \text{otherwise} \end{cases} \qquad (12.11)$$

To obtain the estimations of w and b, Eq. (12.10) is transformed to the primal function given by Eq. (12.12) by introducing the positive slack variables ζ_i and ζ_i^* as follows:

Minimise $R_{\text{SVMs}}(w, \zeta^{(*)}) = \dfrac{1}{2}\|w\|^2 + C\sum_{i=1}^{n}(\zeta_i, \zeta_i^*)$

Subjected to $d_i - w\phi(x_i) - b_i \le \varepsilon + \zeta_i$ (12.12)

$w\phi(x_i) + b_i - d_i \le \varepsilon + \zeta_i^*, \quad \zeta_i^{(*)} \ge 0$

Finally, by introducing Lagrange multipliers and exploiting the optimality constraints, the decision function given by Eq. (12.9) has the following explicit form:

$$f(x, a_i, a_i^*) = \sum_{i=1}^{n}(a_i - a_i^*)K(x, x_i) + b \qquad (12.13)$$

In Eq. (12.13), a_i and a_i^* are the Lagrange multipliers. They satisfy the equalities $a_i * a_i^* = 0$, $a_i \ge 0$ and $a_i^* \ge 0$, where $i = 1; :::; n$, and are obtained by maximising the dual function of Eq. (12.12), which has the following form:

$$R(a_i, a_i^*) = \sum_{i=1}^{n} d_i(a_i - a_i^*) - \varepsilon\sum_{i=1}^{n}(a_i - a_i^*)$$

$$-\dfrac{1}{2}\sum_{i=1}^{n}\sum_{j=1}^{n}(a_i - a_i^*)(a_j - a_j^*)K(x_i, x_j) \qquad (12.14)$$

With the constraints,

$$\sum_{i=1}^{n}(a_i - a_i^*) = 0,$$

$0 \le a_i \le C, \quad i = 1, 2, \ldots, n,$

$0 \le a_i^* \le C, \quad i = 1, 2, \ldots, n$

Based on the Karush–Kuhn–Tucker (KKT) conditions of quadratic programming, only a certain number of coefficients $(a_i - a_i^*)$ in Eq. (12.13) will assume non-zero values. The data points associated with them have approximation errors equal to or larger than ε and are referred to as support vectors. These are the data points lying on or outside the ε-bound of the decision function.

$K(x_i; x_j)$ is defined as the kernel function (Gunn, 1998). The value of the kernel is equal to the inner product of two vectors X_i and X_j in the feature space $\varphi(x_i)$ and $\varphi(x_j)$, i.e., $K(x_i; x_j) = \varphi(x_i) * \varphi(x_j)$. Some of the available kernel functions are as follows:

Linear kernel: $K = X * X^T$

Sigmoid kernel: $K = \tanh(P_1 * X * X^T / \text{length}(X) + P_2)$

where, P_1 and P_2 are user-specified parameters. The final SVR model can be expressed in the following form:

$$X_{k+1} = \beta K + b$$

where, β is the difference of Lagrange multipliers, K is the kernel function and b is the bias term.

12.5 VERIFICATION AND VALIDATION

The process of verification involves testing the model using the same input sequence which was used for developing the model, and the process of validation involves testing the model using a new input sequence. The performance of the model is then numerically expressed in terms of error between the actual and the model output or using the correlation value between the actual and the model outputs.

12.6 THE NORMAL DISTRIBUTION

A random variable X is $N(\mu, \sigma^2)$ if it is normally distributed with mean μ and variance σ^2. Then, the density of X is $\dfrac{1}{\sigma\sqrt{2\pi}} \exp\left[-\dfrac{1}{2}\dfrac{(x-\mu)^2}{\sigma^2}\right]$. If X is $N(\mu, \sigma^2)$, then $\dfrac{(x-\mu)}{\sigma}$ is $N(0, 1)$, i.e., $\dfrac{(x-\mu)}{\sigma}$ is standard normal and the standard normal density is $\phi(x) = \dfrac{1}{\sqrt{2\pi}} \exp\left[-\dfrac{1}{2}x^2\right]$.

12.7 MATHEMATICAL MODELLING USING MAXIMUM ENTROPY PRINCIPLE (MEP)

All probability distributions have some uncertainty associated with them. Entropy provides a quantitative measure of this uncertainty. By the principle of maximum entropy, we try to find a probability distribution which has maximum entropy or uncertainty out of all other distributions which have prescribed moments. In other words, firstly, we find a family of probability distributions whose members are consistent with the given information, and then, from this family, we choose that distribution whose entropy is greater than the entropy of every other member of this family.

12.8 ENTROPY MEASURES

1. Gibbs entropy: The Gibbs entropy (S) of a collection of particles, with i as a discrete set of microstates and P_i as its probability of occurrence during the system's fluctuations, is given by

$$S = -K_b \sum_{i=1}^{n} P_i \ln P_i$$

where, K_b is the Boltzmann constant.

2. Shannon's entropy: Consider a random sequence $\{x_1, x_2, ..., x_n\}$ of successive observations on the variable X, of a random experiment, and assume that the probability associated with the value x_i is P_i.

$$P(X = x_i) = P_i, \quad i = 1, 2, ..., n$$

Now, defining an ensemble $\{(x_1, P_1), (x_2, P_2), ..., (x_n, P_n)\}$ with $\Sigma P_i = 1$, the information associated with the outcome $\{X = x_i\}$ is given by

$$h(P_i) = -\log P_i$$

The Shannon's entropy $H(X)$ of the ensemble is given by

$$H(X) \equiv H_n(P) = H_n(P_1, P_2, ..., P_n) = -C \sum_{i=1}^{n} P_i \log P_i$$

where, C is an arbitrary positive constant.

The entropy $H_n(P) = \sum_{i=1}^{n} P_i h(P_i)$ is the average information. Since X is a random variable with probability mass function P_X, $H(X)$ can be expressed as $H(X) = E[-\log P_X]$, where E is the expectation.

EXAMPLE 12.1 Calculate the Shannon's entropy of X, where $X = [1\ 2\ 2\ 3\ 3\ 3\ 4\ 4\ 4\ 4]$.

Solution: Here, the total number of elements, $N = 10$. Computing the frequency of each value in the set and the probabilities,

Value	Frequency (f)	$P_i = f/N$
1	1	0.1
2	2	0.2
3	3	0.3
4	4	0.4

The Shannon's entropy $H(X)$ is

$H(X) = -[(0.1 \log_2 0.1) + (0.2 \log_2 0.2) + (0.3 \log_2 0.3) + (0.4 \log_2 0.4)]$
$H(X) = -[(-0.332) + (-0.464) + (-0.521) + (-0.529)]$
$H(X) = -[-1.84644]$
$H(X) = 1.84644$

EXAMPLE 12.2 Compute the Shannon's entropy for $X = [1\ 2\ 1\ 3\ 2\ 1\ 4\ 6]$.

Solution: Here, the total number of elements, $N = 8$. Calculating the probabilities,

Value	Frequency (f)	$P_i = f/N$
1	3	0.375
2	2	0.250
3	1	0.125
4	1	0.125
6	1	0.125

The Shannon's entropy $H(X)$ is

$H(X) = -[(0.375 \log_2 0.375) + (0.25 \log_2 0.25) + (0.125 \log_2 0.125) +$
$\quad (0.125 \log_2 0.125) + (0.125 \log_2 0.125)]$
$H(X) = -[(-0.531) + (-0.5) + (-0.375) + (-0.375) + (-0.375)]$
$H(X) = -[-2.15564]$
$H(X) = 2.15564$

3. Joint entropy: Consider a probabilistic system which is described in terms of two random variables X and Y, the joint entropy of X and Y is

$$H(X, Y) = -\sum_{i=1}^{n} \sum_{j=1}^{m} P_{ij} \log P_{ij}$$

where, $P_{ij} = P(x_i, y_i) = P(X = x_i, Y = y_i)$, $i = 1, 2, ..., n$ and $j = 1, 2, ..., m$ is the joint probability mass function.

4. Weighted entropy: If we want to quantify the amount of information supplied by a probabilistic experiment when the experiment is characterised by both the objective probabilities and weights associated with the events, then we characterise each event x_i by $\{P_i, w_i\}$, $i = 1, ..., n$, with $\sum_{i=1}^{n} P_i = 1$ and $w_i \geq 0$. Here, P_i is the probability of the event x_i and w_i is the weight assigned to it. Now, the weighted entropy is defined as

$$H_w(w_1, ..., w_n; P_1, ..., P_n) = -\sum_{k=1}^{n} w_k P_k \log P_k$$

5. Differential or continuous (Boltzmann) entropy: The entropy of a continuous random variable X with probability density function $f(x)$ is defined as

$$h(X) = -\int_R f(x) \log f(x) dx$$

R is the support set where $f(x) > 0$.

The differential entropy of a set $X_1, X_2, \ldots, X_n$ of random variables with the density function $f(x_1, x_2, \ldots, x_n)$ is defined as

$$h(X_1, X_2, \ldots, X_n) = -\int f(x_1, x_2, \ldots, x_n) \log f(x_1, x_2, \ldots, x_n) dx_1 dx_2 \ldots dx_n$$

EXAMPLE 12.3 What is the differential entropy of a random variable X distributed over a uniform distribution $f(x) = \dfrac{1}{a}, 0 \leq x \leq a$?

Solution: The differential entropy of X is

$$h(X) = -\int_b^a \frac{1}{a} \log \frac{1}{a} dx = \log a$$

EXAMPLE 12.4 If $X \sim N(0, \sigma^2)$ with the probability density function $f(x) = \dfrac{1}{\sqrt{2\pi\sigma^2}} e^{-\frac{x^2}{2\sigma^2}}$, find the differential entropy of X.

Solution:
$$h(X) = -\int f(x) \log f(x) dx$$

$$= -\int f(x) \left(\log \frac{1}{\sqrt{2\pi\sigma^2}} - \frac{x^2}{2\sigma^2} \log e \right) dx$$

$$= \frac{1}{2} \log (2\pi\sigma^2) + \frac{\log e}{2\sigma^2} E_f[X^2]$$

$$= \frac{1}{2} \log (2\pi e \sigma^2)$$

6. Tsallis entropy: The Tsallis entropy is a generalised entropy that has found its applications in several fields of science and engineering. Given a set of probabilities $\{P_i\}$ and q be a real number, the Tsallis entropy in a discrete system is defined as

$$S_q^{(d)} = \frac{1 - \sum_{i=1}^n P_i^q}{q - 1}$$

where, $\{P_i\}_{i=1}^n$ is a probability distribution.

The Tsallis entropy in a continuous system is defined as

$$S_q^{(c)} = \frac{1 - \int f(x)^q\, dx}{q-1}$$

where, f is a probability density function.

The Tsallis entropy converges to Shannon's entropy when $q \to 1$.

$$\lim_{q \to 1} S_q^{(d)} = -\sum_{i=1}^{n} P_i \ln P_i$$

and

$$\lim_{q \to 1} S_q^{(c)} = -\int f(x) \ln f(x)\, dx$$

The maximum entropy principle for Tsallis entropy $S_q^{(c)}$ under the constraints

$$\int f(x)\, dx = 1$$

and

$$\frac{\int x^2 f(x)^q\, dx}{\int f(y)^q\, dy} = \sigma^2$$

yields the q-Gaussian probability density function

$$f(x) = \frac{\exp_q(-\beta_q x^2)}{\int \exp_q(-\beta_q y^2)\, dy} \propto [1 + (1-q)(-\beta_q x^2)]^{\frac{1}{1-q}}$$

where, $\exp_q(x) = \begin{cases} [1 + (1-q)x]^{\frac{1}{1-q}} & \text{if } 1 + (1-q)x > 0, (x \in \mathbb{R}) \\ 0 & \text{otherwise} \end{cases}$

and β_q is a positive constant related to σ and q. For $q \to 1$, the q-Gaussian distribution recovers a usual Gaussian distribution.

Further, the q-logarithm function can be expressed as

$$\ln_q x = \frac{x^{1-q} - 1}{1-q}, \quad (x \geq 0, q \in \mathbb{R})$$

whose inverse function is the q-exponential function. Using the q-logarithm function, the Tsallis entropy $S_q^{(c)}$ can be given as

$$S_q^{(c)} = -\int f(x)^q \ln_q f(x)\, dx$$

which converges to Shannon's entropy when $q \to 1$.

System Identification

7. Rényi entropy: This measure is used widely in the study of quantum systems, quantum measurements and quantum statistical mechanics. The Rényi entropy of order α is defined as

$$H_\alpha = \frac{1}{1-\alpha} \ln\left(\sum_{i=1}^{n} P_i^\alpha\right), \quad \alpha \geq 0$$

which is the measure of information of order α associated with the probability distribution $\{P_i\}_{i=1}^{n}$. The Rényi entropy is viewed as a measure of uncertainity, since the uncertainity is nothing but the missing information. The Rényi entropy is a decreasing function of α and when $\alpha \to 1$, the Rényi entropy is equal to Shannon's entropy.

$$\lim_{\alpha \to 1} H_\alpha = -\sum P_i \ln P_i$$

8. Relative entropy: The relative entropy (or Kullback–Leibler distance) $D(f \parallel g)$ between two densities f and g is defined as

$$D(f \parallel g) = \int f \log \frac{f}{g}$$

9. Mutual information: The mutual information $I(X; Y)$ between the two continuous random variables X and Y with the joint density function $f(x, y)$ is defined as

$$I(X; Y) = \int f(x, y) \log \frac{f(x, y)}{f(x) f(y)} \, dx \, dy$$

10. von Neumann entropy: Let ρ_x be a set of quantum states, each occurring with a specified probability P_x. Then, the density matrix ρ is given by

$$\rho = \sum_x P_x \rho_x$$

For a quantum mechanical system described by a density matrix ρ, the von Neumann entropy is given by

$$S(\rho) = -\text{Trace}(\rho \log \rho)$$

12.9 PROCEDURE FOR SYSTEM IDENTIFICATION USING MAXIMUM ENTROPY PRINCIPLE (MEP)

We now discuss a grey-box type of approach for obtaining the mathematical model of a system using MEP. The first step involved in this procedure is to obtain a general functional form or ordinary differential equation representing the relationships involved in the system. Now, since the mathematical structure of the system is known, we try to obtain information about all the possible ranges of the unknown parameters of the model. The next step is to numerically simulate the model based on uniform distributions for parameters and initial conditions. Now, we compare the results with the observation. Again, we simulate the model on the basis of new maximum entropy distributions for parameters, and again, we compare the results with the observations. This process is repeated until we get satisfactory results. This procedure is efficient for ecological modelling and biological modelling (Kapur, 1988).

For example, consider the logistic model for population growth.

$$\frac{dx}{dt} = ax - bx^2$$

where, $x(t)$ is the population at time t, and a and b are the parameters. The identification problem involves fitting the measured values of x to the model, and thereby, finding the values of a and b so that the model output is close to the measured data.

In most cases, we may have only partial knowledge about the parameters. In this case, the most unbiased distribution for each parameter is the uniform distribution. Also, we can assume that the initial conditions follow a uniform distribution and we choose some random parameter values and initial conditions on this basis. Now, if we can prescribe the mean and variances, the maximum entropy distribution will be a truncated normal distribution and we can draw random samples from these distributions.

For example, consider Figure 12.17. For the measured or observed data points, we can fit several models that pass through all the points. However, the curve which maximizes the entropy may provide more information about the system.

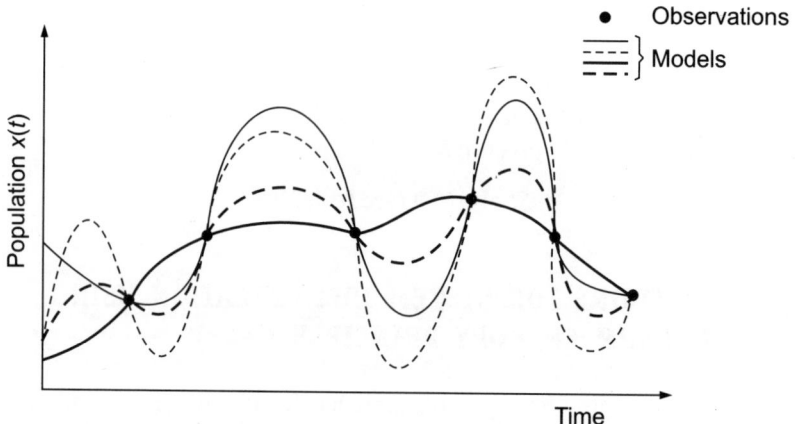

FIGURE 12.17 Modelling using maximum entropy principle.

EXERCISES

1. Explain the steps involved in identification of non-linear systems using a suitable flow chart.
2. State the principle of optimal design of experiments.
3. List out any two persistently exciting signals.
4. Explain the procedure for identification of the Hammerstein–Weiner model.
5. Approximate the Lotka–Volterra model using artificial neural networks, for the Lotka–Volterra model parameters: $a = 0.3$, $b = 0.9$, $c = 0.5$ and $d = 0.7$. Use MATLAB to obtain the ANN model.
6. Calculate the Shannon's entropy of X, where

$$X = \begin{pmatrix} 1 \\ 1.1 \\ 2.1 \\ 2.2 \\ 2.3 \\ 2.4 \\ 3.3 \\ 3.4 \\ 3.5 \\ 3.2 \end{pmatrix}$$

7. Explain the relation between entropy and information content.

CHAPTER 13
Parameter Estimation Techniques

Parameter estimation is the method of using measured data to estimate the unknown parameters of a mathematical model. Parameter estimation techniques can be classified into linear methods and non-linear methods (Figure 13.1) applicable to linear and non-linear models, respectively. Further, a *recursive parameter estimation technique* is a technique for estimating the parameters of a time-variant model. These techniques are used for online parameter estimation.

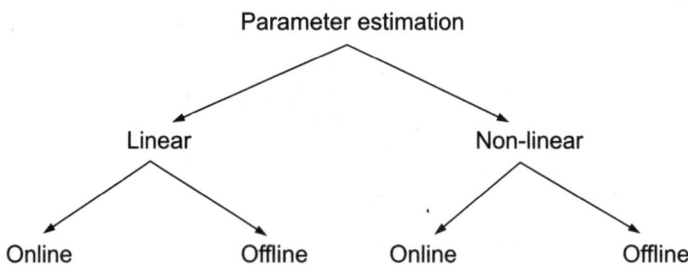

FIGURE 13.1 Classification of parameter estimation techniques.

13.1 ESTIMATION OF PARAMETERS OF LINEAR MODELS: LEAST SQUARES METHOD

The least squares method is a well-known technique for estimating the parameter vector of a linear model by the minimisation of the sum of the squares of the residuals.

Consider that there are k pairs of values for two variables x_i and y_i, and if the variables are related by a linear equation of the form

$$y = xa \qquad (13.1)$$

where, x is known exactly and the measure of y is only approximate (y^*), then

$$y^* = y + \varepsilon_y = xa + \varepsilon_y \qquad (13.2)$$

where, ε_y is the error. The least squares method requires that the estimated value of a (denoted by $\hat{a}$) be chosen, which minimises the least squares objective function J.

$$J = \sum_{i=1}^{k} (x_i \hat{a} - y_i^*)^2 \qquad (13.3)$$

The value of $\hat{a}$ is found by taking the gradient of J with respect to $\hat{a}$ and equating to zero.

$$\frac{\partial J}{\partial \hat{a}} = 2 \sum_{i=1}^{n} x_i (x_i a - y_i^*) = 0 \qquad (13.4)$$

By solving Eq. (13.4), we get the estimate $\hat{a}_k$ that minimises J after k samples are given.

$$\hat{a}_k = \left(\sum_{i=1}^{k} x_i^2 \right)^{-1} \sum_{i=1}^{k} x_i y_i^* \qquad (13.5)$$

$$= P_k b_k$$

where,

$$P_k = \left(\sum_{i=1}^{k} x_i^2 \right)^{-1} \qquad (13.6)$$

$$b_k = \sum_{i=1}^{k} x_i y_i^* \qquad (13.7)$$

For a time-variant model, a recursive form of parameter estimation is required in which the estimate at the previous $(k-1)$th instant is used to find the estimate $\hat{a}$ at kth instant. Now, let us write the parameter estimation equation in a different form.

$$P_k^{-1} = \sum_{i=1}^{k} x_i^2$$

$$= P_{k-1}^{-1} + x_k^2 \qquad (13.8)$$

and

$$b_k = \sum_{i=1}^{k} x_i y_i^*$$

$$= b_{k-1} + x_k y_k^* \qquad (13.9)$$

Rearranging Eq. (13.8),

$$P_{k-1} = P_k + P_k x_k^2 P_{k-1}$$

and

$$P_{k-1} x_k = P_k x_k + P_k x_k^2 P_{k-1}$$
$$= P_k x_k (1 + P_{k-1} x_k^2) \qquad (13.10)$$

or

$$P_{k-1} x_k (1 + P_{k-1} x_k^2)^{-1} = P_k x_k$$

Now, multiplying by $P_{k-1} x_k$ and using Eq. (13.10), we get

$$P_{k-1}^2 x_k^2 (1 + P_{k-1} x_k^2)^{-1} = P_k x_k^2 P_{k-1} = P_{k-1} - P_k$$

and consequently,

$$P_k = P_{k-1} - P_{k-1}^2 x_k^2 (1 + P_{k-1} x_k^2)^{-1} \qquad (13.11)$$

Now, substituting Eq. (13.11) in Eq. (13.5), and using Eq. (13.8), we get

$$\hat{a}_k = [P_{k-1} - P_{k-1}^2 x_k^2 (1 + P_{k-1} x_k^2)^{-1}](b_{k-1} + x_k y_k^*)$$

and

$$\hat{a}_{k-1} = P_{k-1} b_{k-1}$$

which can be expanded to get

$$\hat{a} = \hat{a}_{k-1} - K_k (x_k \hat{a}_{k-1} - y_k^*) \qquad (13.12)$$

where,

$$K_k = P_{k-1} x_k (1 + P_{k-1} x_k^2)^{-1} \qquad (13.13)$$

or

$$K_k = (P_k P_k^{-1}) P_{k-1} x_k (1 + P_{k-1} x_k^2)^{-1} \qquad (13.14)$$

Now, substituting Eq. (13.8) in Eq. (13.14), we get

$$K_k = P_k (P_{k-1}^{-1} + x_k^2) P_{k-1} x_k (1 + P_{k-1} x_k^2)^{-1}$$
$$= P_k (x_k + x_k^2 P_{k-1} x_k)(1 + P_{k-1} x_k^2)^{-1}$$
$$= P_k x_k \qquad (13.15)$$

The estimation algorithm is given by Eqs. (13.11), (13.12), (13.13) and (13.15). The starting values of $\hat{a}$ and P must be chosen appropriately. One approach is to allow $\hat{a}_0$ to have a small initial value and set P_0 to a large positive number.

Now, consider the problem of estimating multiple parameters of a linear system of the form

$$y = a_1 x_1 + a_2 x_2 + \dots + a_n x_n \qquad (13.16)$$

where, a_j is the set of unknown parameters to be estimated from the measurement y^* of y and x_j are exactly known quantities. In this case, the minimisation of the least squares function

$$J = \sum_{i=1}^{k} \left(\sum_{j=1}^{n} x_{ji} \hat{a}_j - y_i^* \right)^2 \quad (13.17)$$

requires that all the partial derivatives of J with respect to each of the parameter estimates $\hat{a}_j$ should be simultaneously equal to zero.

Writing Eq. (13.16) in matrix form,

$$y = X^T A \quad (13.18)$$

where, $X^T = [x_1, x_2, ..., x_n]$

and $\quad A[a_1, a_2, ..., a_n]$

Now, Eq. (13.17) can be defined as

$$J = \sum_{i=1}^{k} (X_i^T \hat{A} - y_i^*)^2 \quad (13.19)$$

where $\quad y_i^* = X_i^T A + \varepsilon_y \quad (13.20)$

Now, the gradient equation can be written as

$$\nabla_{\hat{A}}(J) = \left(\sum_{i=1}^{k} X_i X_i^T \right) \hat{A} - \sum_{i=1}^{k} X_i y_i^* = 0 \quad (13.21)$$

When $\nabla_{\hat{A}}(J)$ denotes the gradient of J with respect to all the elements of $\hat{A}$, the solution of Eq. (13.21) has the form

$$\hat{A}_k = P_k B_k \quad (13.22)$$

where, P_k is an $n \times n$ matrix and B_k is an $n \times 1$ vector. The scalar relations given by Eqs. (13.8) and (13.9) are now written in matrix vector form

$$P_k^{-1} = P_{k-1}^{-1} + X_k X_k^T \quad (13.23)$$

and $\quad B_k = B_{k-1} + X_k y_k^* \quad (13.24)$

For obtaining the recursive form of Eq. (13.22), we pre-multiply Eq. (13.23) by P_k, and then, post-multiply by P_{k-1} to get

$$P_{k-1} = P_k + P_k X_k X_k^T P_{k-1} \quad (13.25)$$

Now, post-multiplying Eq. (13.25) by X_k, we get:

$$P_{k-1} X_k = P_k X_k + P_k X_k X_k^T P_{k-1} X_k$$

$$= P_k X_k (1 + X_k^T P_{k-1} X_k) \quad (13.26)$$

Now, post-multiplying Eq. (13.26) by $(1 + X_k^T P_{k-1} X_k)^{-1} X_k^T P_{k-1}$, we get

$$P_{k-1} X_k [1 + X_k^T P_{k-1} X_k]^{-1} X_k^T P_{k-1} = P_k X_k X_k^T P_{k-1} \quad (13.27)$$

Now, using Eqs. (13.25) and (13.27), we get

$$P_k = P_{k-1} - P_{k-1}X_k[1 + X_k^T P_{k-1} X_k]^{-1} X_k^T P_{k-1} \qquad (13.28)$$

Just like the single-parameter case, the equivalent recursive equation is obtained using Eqs. (13.28) and (13.22) as

$$\hat{A}_k = \hat{A}_{k-1} - P_{k-1}X_k(1 + X_k^T P_{k-1} X_k)^{-1}(X_k^T \hat{A}_{k-1} - y_k^*) \qquad (13.29)$$

or

$$\hat{A}_k = \hat{A}_{k-1} - P_k(X_k X_k^T \hat{A}_{k-1} - X_k y_k^*) \qquad (13.30)$$

Once again, it is necessary to specify starting values of $\hat{A}$ and P.

Now, consider a case where we would like to perform an offline estimation of the mass M, spring constant K and dashpot coefficient B for the system shown in Figure 13.2, using the least square approach.

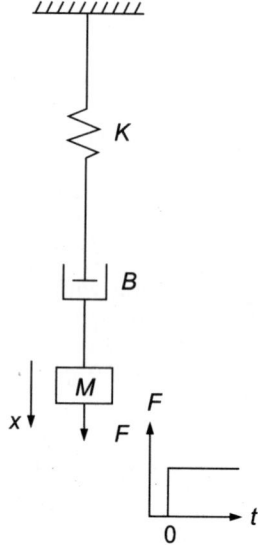

FIGURE 13.2 The mass–spring–dashpot system excited by an input force.

The governing equation of the system can be expressed as

$$M\frac{d^2 x}{dt^2} + B\frac{dx}{dt} + Kx = F$$

where, F is the applied force, $\dfrac{d^2 x}{dt^2}$ is the acceleration, $\dfrac{dx}{dt}$ is the velocity and x is the displacement.

Now, consider that for an applied dynamic force, the acceleration is measured using an accelerometer. It is now possible to integrate the acceleration to obtain velocity, which is further integrated to obtain displacement.

Now, suppose if we want to estimate the parameters $\theta = [M\ B\ K]$ of the system from the applied force and measured acceleration, we can use the least squares approach in the following way:

Let
$$\theta^T = [M\ B\ K]$$

$$\phi = \left[\frac{d^2x}{dt^2}\ \frac{dx}{dt}\ x\right],$$

$$Y = F$$

where, the values of ϕ and Y are available from measurement, due to which the system can be represented in matrix vector form as

$$\phi\theta = Y$$

Now, θ can be estimated using the least squares formula

$$\theta = [\phi^T\phi]^{-1}\phi^T Y$$

The advantage of this method is that if there are multiple measurement points, the matrix $[\phi^T\phi]$ is always a square matrix and investing the square matrix is easy.

Now, consider another example where we would like to fit an exponentially increasing population shown in Figure 13.3 to a linear first order differential equation of the form

$$\frac{dx(t)}{dt} = Ax(t) \tag{13.31}$$

x is the measured population value given as a function of time in a tabulated form and A is a parameter. The analytical solution of this equation is given as

$$x(t) = x_0\, e^{At} \tag{13.32}$$

Now, note that the least squares approach is applicable only for linear algebraic systems. However, in this case, Eq. (13.32) is a non-linear equation with an exponential type of non-linearity. Hence, in order to use the least squares approach, few manipulations of Eq. (13.32) must be performed. Since x_0 and $x(t)$ are measured, let us write Eq. (13.32) as

$$\frac{x(t)}{x_0} = e^{At}$$

Now, taking log on both sides, we get

$$\log\left[\frac{x(t)}{x_0}\right] = At$$

or

$$At = \log\left[\frac{x(t)}{x_0}\right] \tag{13.33}$$

Since the values of t, $x(t)$ and x_0 are available, $\log\left[\dfrac{x(t)}{x_0}\right]$ is just a number, and now, Eq. (13.33) is of the form

$$\phi\theta = Y$$

and least squares method can be applied to obtain the value of A.

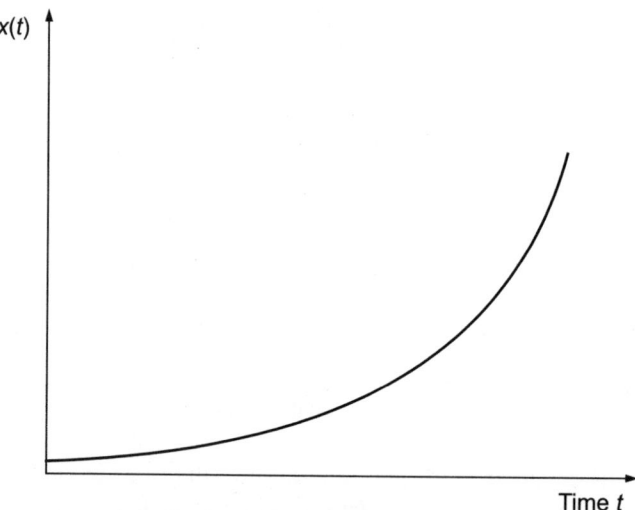

FIGURE 13.3 Measured population x shown as a function of time.

13.2 PARAMETER ESTIMATION METHODS FOR NON-LINEAR MODELS

The methods of least squares and recursive least squares are applicable to linear models only. However, for non-linear models, advanced methods are required. In this section, we discuss two analytical optimisation techniques, namely, the extended Kalman filter (EKF) (Hoshiya & Saito, 1984; Einicke & White, 1999) and unscented Kalman filter (UKF) (Julier & Uhlmann 2004; Wan & Van Der Merwe 2001; Haykin, 2004) algorithms for the estimation of non-linear model parameters. Also, details regarding dynamic programming as well as maximum likelihood estimation are provided. Further, we discuss three different intelligent optimisation techniques based on evolutionary computation and swarm intelligence, namely, genetic algorithm (GA), particle swarm optimisation (PSO) and bacterial foraging optimisation (BFO), respectively, for the estimation of model parameters.

13.2.1 THE EXTENDED KALMAN FILTER

The EKF is used in cases where the dynamic model is non-linear. In the extended Kalman filter, the estimation around the current estimate is linearised using the partial derivatives of the process and measurement functions for computing the estimates. The EKF is actually a state estimator, which can be used for online estimation of the parameters of a differential equation or difference equation when few state measurements are available.

Let the state vector be $x \in \mathbb{R}^n$ and the process is governed by the non-linear difference equation described by Eq. (13.34) and let the measurement be $z \in \mathbb{R}^m$.

$$x_k = f(x_{k-1}, w_{k-1}) \tag{13.34}$$

$$z_k = h(x_k, v_k) \tag{13.35}$$

where, the random variables w_k and v_k represent the process and measurement noise. The non-linear function f in Eq. (13.34) relates the state at the previous time step $k - 1$ to the state at the current time step k.

The probability distributions of w and v are $p(w) \sim N(0, Q)$ and $p(v) \sim N(0, R)$. In practice, one does not know the individual values of the noise and at each time step. However, one can approximate the state and measurement vector without them as

$$\tilde{x}_k = f(\hat{x}_{k-1}, 0) \tag{13.36}$$

$$\tilde{z}_k = h(\tilde{x}_k, 0) \tag{13.37}$$

where, $\hat{x}_k$ is a posteriori estimate of the state (from a previous time step k). To estimate a process with non-linear relationships, new governing equations that linearise an estimate about Eqs. (13.36) and (13.37), are written.

$$x_k \approx \tilde{x}_k + A(x_{k-1} - \hat{x}_{k-1}) + W w_{k-1} \tag{13.38}$$

$$z_k \approx \tilde{z}_k + H(x_k - \tilde{x}_k) + V v_k \tag{13.39}$$

where, x_k and z_k are the actual state and measurement vectors, $\tilde{x}_k$ and $\tilde{z}_k$ are the approximate state and measurement vectors from Eqs. (13.36) and (13.37), $\hat{x}_k$ is a posteriori estimate of the state at step k, the random variables w_k and v_k represent the process and measurement noise.

A is the Jacobian matrix of partial derivatives of $f(.)$ with respect to x, i.e.,

$$A_{[i,j]} = \frac{\partial f_{[i]}}{\partial x_{[j]}} (\hat{x}_{k-1}, 0)$$

W is the Jacobian matrix of partial derivatives of $f(.)$ with respect to w, i.e.,

$$W_{[i,j]} = \frac{\partial f_{[i]}}{\partial w_{[j]}} (\hat{x}_{k-1}, 0)$$

H is the Jacobian matrix of partial derivatives of $h(.)$ with respect to x, i.e.,

$$H_{[i,j]} = \frac{\partial h_{[i]}}{\partial x_{[j]}} (\tilde{x}_k, 0)$$

V is the Jacobian matrix of partial derivatives of $h(.)$ with respect to v, i.e.,

$$V_{[i,j]} = \frac{\partial h_{[i]}}{\partial v_{[j]}}(\hat{x}_k, 0)$$

Now, the prediction error is defined as $\tilde{e}_{x_k} \equiv x_k - \tilde{x}_k$ and the measurement residual as $\tilde{e}_{z_k} \equiv z_k - \tilde{z}_k$. The governing equations for the error process are written as

$$\tilde{e}_{x_k} \approx A(x_{k-1} - \tilde{x}_{k-1}) + \varepsilon_k \tag{13.40}$$

$$\tilde{e}_{z_k} \approx H\tilde{e}_{x_k} + \eta_k \tag{13.41}$$

$$\hat{x}_k = \tilde{x}_k + \hat{e}_k \tag{13.42}$$

where, ε_k and η_k represent new independent random variables having zero mean and covariance matrices WQW^T and VRV^T. The probability distributions of the random variables are

$$p(\tilde{e}_{x_k}) \sim N(0, E[\tilde{e}_{x_k}\tilde{e}_{x_k}^T]) \tag{13.43}$$

$$p(\varepsilon_k) \sim N(0, WQ_kW^T) \tag{13.44}$$

$$p(\eta_k) \sim N(0, VR_kV^T) \tag{13.45}$$

The Kalman filter equation used to estimate $\hat{e}_k$ is $\hat{e}_k = K_k\tilde{e}_{z_k}$, and hence,

$$\hat{x}_k = \tilde{x}_k + K_k\tilde{e}_{z_k} \tag{13.46}$$

$$= \tilde{x}_k + K_k(z_k - \tilde{z}_k) \tag{13.47}$$

where, K_k is the Kalman gain and is obtained using the derivation of the basic Kalman filter. To summarise the EKF algorithm, the time and measurement update equations are given as
    Time update

$$\hat{x}_k^- = f(\hat{x}_{k-1}, 0) \tag{13.48}$$

$$P_k^- = A_k P_{k-1} A_k^T + W_k Q_{k-1} W_k^T \tag{13.49}$$

Measurement update

$$K_k = P_k^- H_k^T (H_k P_k^- H_k^T + V_k R_k V_k^T)^{-1} \tag{13.50}$$

$$\hat{x}_k = \hat{x}_k^- + K_k(z_k - h(\hat{x}_k^-, 0)) \tag{13.51}$$

$$P_k = (I - K_k H_k)P_k^- \tag{13.52}$$

13.2.2 The Unscented Kalman Filter

The unscented Kalman filter is founded on the intuition that it is easier to approximate a probability distribution than to approximate an arbitrary non-linear function. The sigma points are chosen so that their mean and covariance are exactly x^a_{k-1} and P_{k-1}. Each sigma point is then propagated through the non-linearity, yielding in the end a cloud of transformed points. The new estimated mean and covariance are then computed based on their statistics. This process is called *unscented transformation*. The unscented transformation is a method for calculating the statistics of a random variable, which undergoes a non-linear transformation.

Consider the following non-linear system described by the difference equation and the observation model with additive noise.

$$x_k = f(x_{k-1}) + w_{k-1} \qquad (13.53)$$

$$z_k = h(x_k) + v_k \qquad (13.54)$$

The initial state x_0 is a random vector with known mean $\mu_0 = E[x_0]$ and covariance $P_0 = E[(x_0 - \mu_0)(x_0 - \mu_0)^T]$.

Selection of Sigma Points

Let X_{k-1} be a set of $2n + 1$ sigma points (where n is the dimension of the state space) and their associated weights

$$X_{k-1} = \{(x^j_{k-1}, W^j)| j = 0, ..., 2n\} \qquad (13.55)$$

Consider the following selection of sigma points, which that incorporates higher order information in the selected points:

$$x^0_{k-1} = x^a_{k-1} \qquad (13.56)$$

$$-1 < W^0 < 1$$

$$x^i_{k-1} = x^a_{k-1} + \left(\sqrt{\frac{n}{1 - W^0} P_{k-1}}\right)_i \quad \text{for all } i = 1 \ldots n \qquad (13.57)$$

$$x^{i+n}_{k-1} = x^a_{k-1} - \left(\sqrt{\frac{n}{1 - W^0} P_{k-1}}\right)_i \quad \text{for all } i = 1 \ldots n \qquad (13.58)$$

$$W^j = \frac{1 - W^0}{2n} \quad \text{for all } j = 1 \ldots 2n \qquad (13.59)$$

where, the weights must obey the following condition:

$$\sum_{j=0}^{2n} W^j = 1 \qquad (13.60)$$

and $\left(\sqrt{\frac{n}{1 - W^0} P_{k-1}}\right)_i$ is the row or column of the matrix square root of $\frac{n}{1 - W^0} P_{k-1}$.

W^0 **controls the position of sigma points:** $W^0 \geq 0$ points tend to move further from the origin, $W^0 \leq 0$ points tend to be closer to the origin.

Model Forecast Step

Each sigma point is propagated through the non-linear process model.

$$x_k^{f,j} = f(x_{k-1}^j) \tag{13.61}$$

The transformed points are used to compute the mean and covariance of the forecast value of x_k.

$$x_k^f = \sum_{j=0}^{2n} W^j x_k^{f,j} \tag{13.62}$$

$$P_k^f = \sum_{j=0}^{2n} W^j (x_k^{f,j} - x_k^f)(x_k^{f,j} - x_k^f)^T + Q_{k-1} \tag{13.63}$$

where, Q_{k-1} is the process noise covariance matrix. We propagate the sigma points through the non-linear observation model.

$$z_{k-1}^{f,j} = h(x_{k-1}^j) \tag{13.64}$$

With the resulted transformed observations, their mean and covariance (innovation covariance) are computed.

$$z_{k-1}^f = \sum_{j=0}^{2n} W^j z_{k-1}^{f,j} \tag{13.65}$$

$$\text{Cov}(\tilde{z}_{k-1}^f) = \sum_{j=0}^{2n} W^j (z_{k-1}^{f,j} - z_{k-1}^f)(z_{k-1}^{f,j} - z_{k-1}^f)^T + R_k \tag{13.66}$$

where, R_k is the measurement noise covariance matrix. The cross covariance between $\tilde{x}_k^f$ and $\tilde{z}_{k-1}^f$ is

$$\text{Cov}(\tilde{x}_k^f, \tilde{z}_{k-1}^f) = \sum_{j=0}^{2n} W^j (x_k^{f,j} - x_k^f)(z_{k-1}^{f,j} - z_{k-1}^f)^T \tag{13.67}$$

Data assimilation step

Assume that the estimate has the following form:

$$x_k^a = x_k^f + K_k(z_k - z_{k-1}^f) \tag{13.68}$$

The gain K_k is given by

$$K_k = \text{Cov}(\tilde{x}_k^f, \tilde{z}_{k-1}^f)\, \text{Cov}^{-1}(\tilde{z}_{k-1}^f) \tag{13.69}$$

The posterior covariance is updated using

$$P_k = P_k^f - K_k \text{Cov}(\tilde{z}_{k-1}^f) K_k^T \tag{13.70}$$

The parameter estimation problem involves estimating a set of unknown parameters in a non-linear mapping $y_k = G(x_k, w)$, where w corresponds to the set of unknown parameters. The parameter set can even be the weights of a neural network if the system is modelled by means of an artificial neural network. The unscented Kalman filter can be used to estimate the parameters of the model using the following set of equations.

1. **Initialisation:**

$$\hat{w}_0 = E[w] \tag{13.71}$$

$$Pw_0 = E[(w - \hat{w}_0)(w - \hat{w}_0)^T] \tag{13.72}$$

For $k \in \{1, ..., \infty\}$,

2. **Time update and sigma point calculation:**

$$\hat{w}_k^- = \hat{w}_{k-1} \tag{13.73}$$

$$P_{wk}^- = P_{wk-1} + R_{k-1}^r \tag{13.74}$$

$$W_{k|k-1} = \left[\hat{w}_k^- \quad \hat{w}_k^- + \gamma\sqrt{P_{w_k}^-} \quad \hat{w}_k^- - \gamma\sqrt{P_{w_k}^-} \right] \tag{13.75}$$

$$D_{k|k-1} = G[x_k, W_{k|k-1}] \tag{13.76}$$

$$\hat{d}_k = G(x_k, \hat{w}_k^-) \tag{13.77}$$

3. **Measurement update equations:**

$$P_{\tilde{d}_k \tilde{d}_k} = \sum_{i=0}^{2L} W_i^{(c)} [D_{i,k|k-1} - \hat{d}_k][D_{i,k|k-1} - \hat{d}_k]^T + R_k^e \tag{13.78}$$

$$P_{w_k d_k} = \sum_{i=0}^{2L} W_i^{(c)} [W_{i,k|k-1} - \hat{w}_k^-][D_{i,k|k-1} - \hat{d}_k]^T \tag{13.79}$$

$$K_k = P_{w_k d_k} P_{\tilde{d}_k \tilde{d}_k}^{-1} \tag{13.80}$$

$$\hat{w}_k = \hat{w}_k^- + K_k (d_k - \hat{d}_k) \tag{13.81}$$

$$P_{w_k} = P_{w_k}^- = K_k P_{\tilde{d}_k \tilde{d}_k} K_k^T \tag{13.82}$$

where, $\gamma = \sqrt{(L + \lambda)}$, and λ is composite scaling parameter, L is dimension of the state, R^r and R^e are noise covariances.

The EKF and UKF methods provide a consistent estimate of the model parameters even if some state variables are not available for measurement and the measurement is corrupted by random noises.

The basic idea behind using the EKF and UKF state estimation algorithms for parameter estimation is that the parameters of the system can be considered as states of the system and estimated. Consider the ODE model of a system of the form

$$\frac{dx}{dt} = f(x(t), \theta, t) + w(t)$$
$$z(t) = h(x(t), t) + v(t) \tag{13.83}$$

where, $x(t)$ is the state variable, θ is the model parameter to be estimated, $z(t)$ is the vector of observations, and $w(t)$ and $v(t)$ are the random noises. Now, suppose if we would like to use the EKF algorithm for estimation of θ. Then, we rewrite Eq. (13.83) as follows:

$$\frac{dx}{dt} = f(x(t), \theta(t), t) + w(t)$$
$$\frac{d\theta}{dt} = 0 + w(t) \tag{13.84}$$
$$z(t) = h(x(t), t) + v(t)$$

Now, the augmented state parameter vector is defined as $x_a^T = [x^T, \theta^T]$. The updated equations of EKF and UKF can be applied on the system shown in Eq. (13.84) for the estimation of the model parameters.

13.2.3 Dynamic Programming

The dynamic programming (DP) is a powerful mathematical technique, which can be applied to a variety of modelling problems. Dynamic programming determines the optimum solution of a multivariable problem by decomposing the problem into n stages. Now, each stage comprises a single variable sub-problem. This method reduces the computational complexity, since at each stage, the optimisation problem involves only one variable. A dynamic programming (DP), model is a recursive equation linking the various stages of a problem such that the optimal solution of each stage is also optimal for the whole problem. The DP modelling can be used for a variety of modelling problems such as cargo loading models, shortest path models, models in reliability engineering, inventory modelling, etc. This modelling is based on the *Bellman's principle of optimality,* which states that the future decisions for the remaining stages must constitute an optimal policy regardless of the policy adopted in the previous stages. The dynamic programming can be used efficiently for solving several optimisation problems. To understand the dynamic programming methodology, let us consider the following function maximisation problem.

EXAMPLE 13.1 Maximise the following equation subject to the given constraint:

max $f(y) = (4y_1 - y_1^2) + (4y_2 - 2y_2^2) + (4y_3 - 3y_3^2)$
subject to $y_1 + y_2 + y_3 = 11$, $y_1 \geq 0$, $i = 1, 2, 3$.

Solution:

$$x_3 = y_1 + y_2 + y_3 = 11$$
Let $x_2 = y_1 + y_2 = x_3 - y_3$
$$x_1 = y_1 = x_2 - y_2$$

and hence, $F_3(x_3) = \max_{y_3} \{(4y_3 - 3y_3^2) + F_2(x_2)\}$

$F_2(x_2) = \max_{y_2} \{(4y_2 - 2y_2^2) + F_1(x_1)\}$

$F_1(x_1) = \max_{y_1} \{(4y_1 - y_1^2)\}, y_1 = x_2 - y_2$

$F_2(x_2) = \max_{y_2} \{(4y_2 - 2y_2^2) + 4(x_2 - y_2) - (x_2 - y_2)^2\}$ (i)

whose maximum can be obtained by calculus

$$\frac{d}{dy_2}\{(4y_2 - 2y_2^2) + 4(x_2 - y_2) - (x_2 - y_2)^2\} = 0$$

$$4 - 4y_2 - 4 + 2(x_2 - y_2) = 0$$

$$2x_2 = 6y_2$$

$$y_2 = \frac{x_2}{3} \quad \text{(ii)}$$

and $\quad \dfrac{d^2}{dy_2^2}(4 - 4y_2 - 4 + 2(x_2 - y_2)) = -6 < 0$

So, it has maximum.
Substituting Eq. (ii) in Eq. (i),

$$F_2(x_2) = \left\{\left(4\frac{x_2}{3} - 2\frac{x_2^2}{9}\right) + 4\left(x_2 - \frac{x_2}{3}\right) - \left(x_2 - \frac{x_2}{3}\right)^2\right\}$$

$$= \left(\frac{4x_2}{3} + \frac{8x_2}{3} - \frac{2x_2^2}{9} - \frac{4x_2^2}{9}\right) = \left\{4x_2 - \frac{2x_2^2}{3}\right\}$$

Now, $F_3(x_3) = \max_{y_3}\left\{4y_3 - 3y_3^2 + 4x_2 - \frac{2x_2^2}{3}\right\}, x_2 = x_3 - y_3$

$$= \max_{y_3}\left\{4y_3 - 3y_3^2 + 4x_3 - 4y_3 - \frac{2}{3}(x_3 - y_3)^2\right\} \quad \text{(iii)}$$

$$\frac{d}{dy_3}\left\{-3y_3^2 + 4x_3 - \frac{2}{3}(x_3 - y_3)^2\right\} = 0$$

Parameter Estimation Techniques **215**

$$\left\{-6y_3 + \frac{4}{3}(x_3 - y_3)\right\} = 0$$

$$-18y_3 + 4x_3 - 4y_3 = 0$$

$$y_3 = \frac{2x_3}{11} \qquad \text{(iv)}$$

$$\frac{d^2}{dy_3^2}\left\{-3y_3^2 + 4x_3 - \frac{2}{3}(x_3 - y_3)^2\right\} = -6 - \frac{4}{3} < 0$$

So, it has maximum.

Using Eqs. (iv) and (iii),

$$F_3(x_3) = \left\{-3x_3 + \frac{4}{121}x_3^2 + 4x_3 - \frac{2}{3}\left(x_3 - \frac{2}{11}x_3\right)^2\right\}$$

$$= \left\{4x_3 - \frac{12}{121}x_3^2 - \frac{162}{363}x_3^2\right\} = 4x_3 - \frac{6}{11}x_3^2$$

But $x_3 = 11$.

Hence, $\quad y_3 = \dfrac{22}{11} = 2,\ x_2 = 11 - 2 = 9,\ y_2 = 3$

$$x_1 = 9 - 3 = 6,\ y_1 = 6$$

The solution is $y_1 = 6,\ y_2 = 3,\ y_3 = 2$.

and $\max f(y) = (4y_1 - y_1^2) + (4y_2 - y_2^2) + (4y_3 - y_3^2) = -22$

13.2.4 MAXIMUM LIKELIHOOD (ML) ESTIMATION

The maximum likelihood estimation technique is commonly used for parameter estimation of a statistical distribution. The likelihood function L for a sample of n observations $(x_1, x_2, x_3, ..., x_n)$ is defined by

$$L = f(x_1, \theta) f(x_2, \theta) \cdots f(x_n, \theta) = \prod_1^n f(x_i, \theta)$$

where, θ is the parameter set of a probability density function. The likelihood function is a function of parameters for a given set of measured data. In the maximum likelihood estimation, θ is chosen such that it provides a high probability of obtaining the actual measured data $(x_1, x_2, x_3, ..., x_n)$. The *maximum likelihood estimator* is defined as the value of θ which maximises L.

Mathematically, the maximum likelihood estimator is given by

$$\left.\frac{\partial L}{\partial \theta}\right|_{\theta=\hat{\theta}} = 0$$

or

$$\left.\frac{\partial \ln L}{\partial \theta}\right|_{\theta=\hat{\theta}} = 0$$

EXAMPLE 13.2 Estimate the parameters μ and σ of a normal distribution using the maximum likelihood estimator for n observations of $x = (x_1, x_2, x_3, \ldots, x_n)$.

Solution: The normal distribution is given by

$$f(x, \mu, \sigma) = \left(\frac{1}{\sigma\sqrt{2\pi}}\right) \exp\left(-\frac{1}{2}\left(\frac{x-\mu}{\sigma}\right)^2\right)$$

For n observations of $x = (x_1, x_2, x_3, \ldots, x_n)$, the likelihood function is given by

$$L = \left(\frac{1}{\sigma\sqrt{2\pi}}\right)^n \exp\sum_{i=1}^{n}\left[-\frac{1}{2}\left(\frac{x_i-\mu}{\sigma}\right)^2\right]$$

Taking ln on both sides,

$$\ln L = -n \ln \sigma\sqrt{2\pi} - \frac{1}{2\sigma^2\sqrt{2\pi}} \sum (x_i - \mu)^2$$

Using ML estimation,

$$\frac{\partial \ln L}{\partial \mu} = 0 \Rightarrow \frac{1}{\sigma^2} \sum (x_i - \mu) = 0$$

and

$$\frac{\partial \ln L}{\partial \sigma} = 0 \Rightarrow -\frac{n}{\sigma} - \frac{1}{2}(-2\sigma^{-3}) \sum (x_i - \mu)^2 = 0.$$

Hence, we get

$$\hat{\mu} = \frac{\sum x_i}{n}$$

and

$$\hat{\sigma}^2 = \frac{\sum (x_i - \hat{\mu})^2}{n}$$

EXAMPLE 13.3 Using the maximum likelihood estimator, estimate the parameter λ of an exponential distribution for a set of n observations of x.

Solution: The exponential distribution is defined as

$$f(x, \lambda) = \lambda e^{-\lambda x}$$

For n observations of x, the likelihood function is given by

$$L(\lambda, x_1, x_2, ..., x_n) = \prod_{i=1}^{n} f(x_i)$$

$$L = \lambda^n e^{-\lambda \sum_{i=1}^{n} x_i}$$

$$\ln L = \ln \left(\lambda^n e^{-\lambda \sum_{i=1}^{n} x_i} \right)$$

$$= n \ln(\lambda) - \lambda \sum_{i=1}^{n} x_i$$

Using ML estimator,

$$\frac{\partial \ln L}{\partial \lambda} = 0$$

$$\Rightarrow \frac{n}{\lambda} - \sum_{i=1}^{n} x_i = 0$$

Hence,

$$\hat{\lambda} = \frac{n}{\sum_{i=1}^{n} x_i}$$

13.2.5 ESTIMATION OF MODEL PARAMETERS USING EVOLUTIONARY OPTIMISATION AND SWARM INTELLIGENCE TECHNIQUES

Consider a mathematical model represented by

$$\dot{X} = f(X, \theta)$$

or

$$X(t+1) = f(X(t), \theta)$$

where, X is the set of state variables which have been measured and is available as tabulated values and θ is the set of model parameters, $\theta = [\theta_1, \theta_2, ..., \theta_n]$. Using evolutionary optimisation and swarm intelligence techniques, it is possible to efficiently estimate all the parameters of the model. Before discussing the various evolutionary optimisation and swarm intelligence techniques, we must first learn about the objective function (or cost function or fitness function), which is the key to model parameter estimation using these methods.

An objective function can be defined as a function which must be minimised so that the values of the model parameters can be estimated. The objective function is a function of error and can be a mean square error (MSE) function or a sum squared error (SSE) function or any other custom function based on the error between the model output and the measured values of state variables.

MSE objective function:

$$\min_{\theta} J = \frac{1}{n} \sum_{i=1}^{n} (x_i - \hat{x}_i)^2$$

where, n is the total number of measured samples, x_i denotes the measured values of state variables and $\hat{x}_i$ is the model output.

SSE objective function:

$$\min_{\theta} J = \sum_{i=1}^{n} (x_i - \hat{x}_i)^2$$

Now, the parameter estimation problem can be described as "what should be the value of θ such that J is minimum". The function of the optimisation algorithm is to minimise J and estimate the value of θ.

Evolutionary Algorithms

The genetic algorithm (GA): The genetic algorithm is one of the most popular and flexible algorithms which can be used to solve hard optimisation problems and a variety of other problems in multiple domains. The basic flow chart of the genetic algorithm is presented in Figure 13.4.

The genetic algorithm uses a population of random bit strings which represent a population of potential solutions. If the population size is large, then the diversity of the members of the population is greater and the area searched by the population is larger.

In the next step, the fitness value of each member of the population is computed using the fitness function specified by the user, which is the objective function for parameter estimation problems and each member of the population represents a potential value of the model parameters. The fitness is a measure of how well each potential solution solves the problem. The higher the fitness value, the better is the solution. After computing the fitness of each member, the selection process is carried out. The members of the population are selected using either roulette wheel selection method or an elitist selection method, as shown in Figures 13.5 and 13.6, respectively.

220   Mathematical Modelling of Systems and Analysis

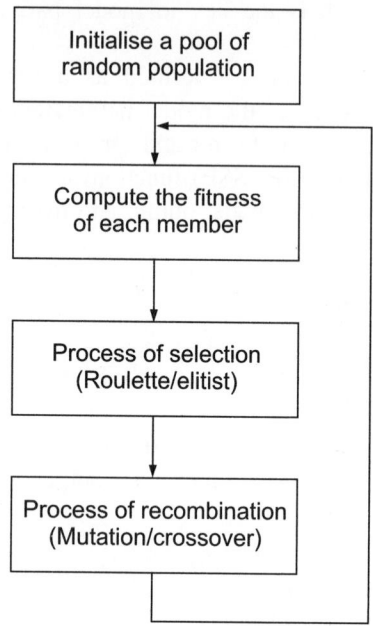

FIGURE 13.4   Flow chart of GA.

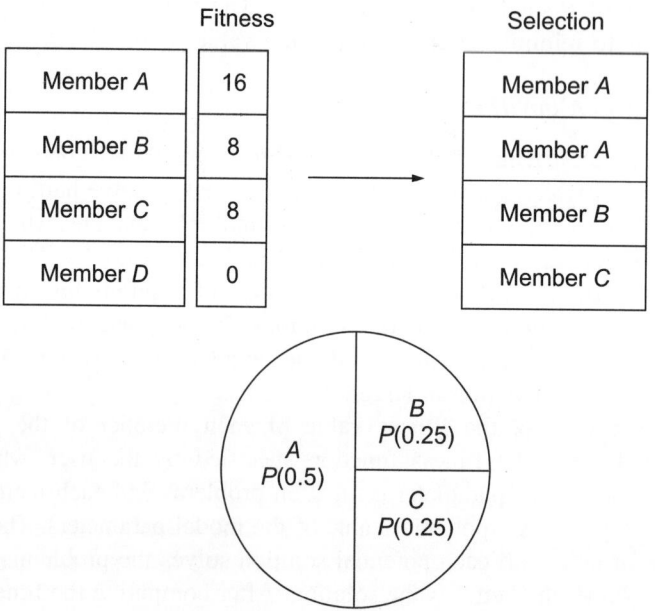

FIGURE 13.5   The roulette wheel selection process.

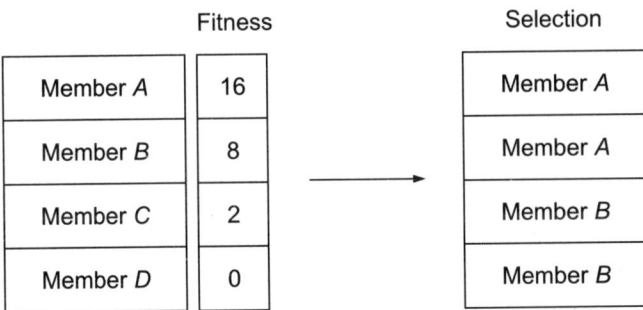

FIGURE 13.6 The elitist selection process.

The roulette wheel selection is a probabilistic algorithm which selects members of the population proportionate with their fitness. In other words, the higher the fitness of the member, the more likely he will be selected.

In elitist selection, the members of the population with higher fitness value are selected and the lesser fit members are forced to die off.

After the selection process, the selected members of the population are given to the next step known as the *recombination step*. In this step, the members are recombined to form new members of the next generation. The parents are selected two at a time from the set of individuals, which are permitted to propagate from the selection process. Two children are created in the new generation from two parents.

The recombination process involves the crossover operation and the mutation operation. The crossover operation can be either a single point operation or a double point operation, as described in Figures 13.7 and 13.8, respectively. Using crossover, the parents are combined by picking a crossover

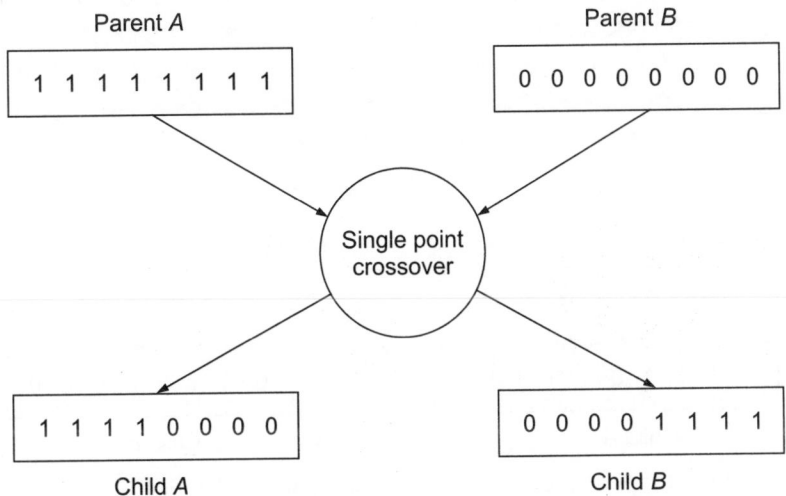

FIGURE 13.7 Single point crossover operation in genetic algorithm.

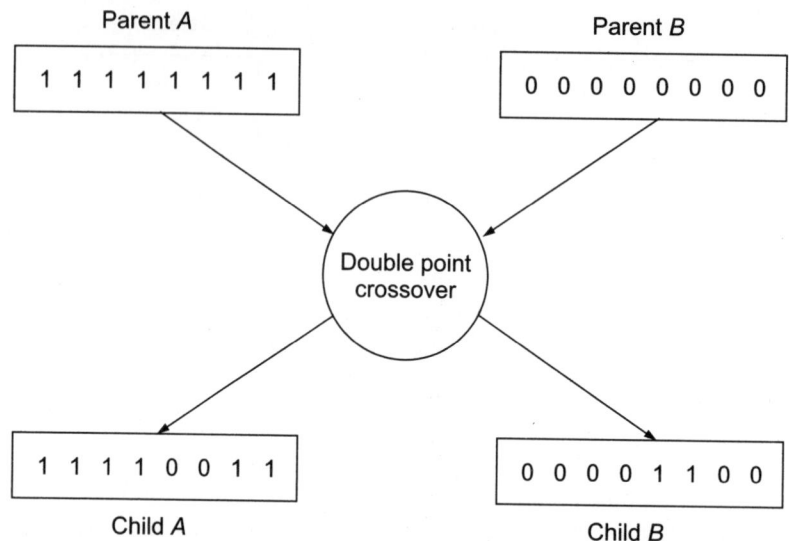

FIGURE 13.8 Double point crossover operation in genetic algorithm.

point and swapping the bits of the parents to produce two children. In the single point scheme, a single crossover point on both the parents is selected and all the bits beyond that point are swapped between the two parents resulting in two children. In the double point scheme, two crossover points are selected on the parent strings and all the bits between these two points are swapped between the parents to produce two children. The mutation operation randomly flips or mutates a bit (Figure 13.9).

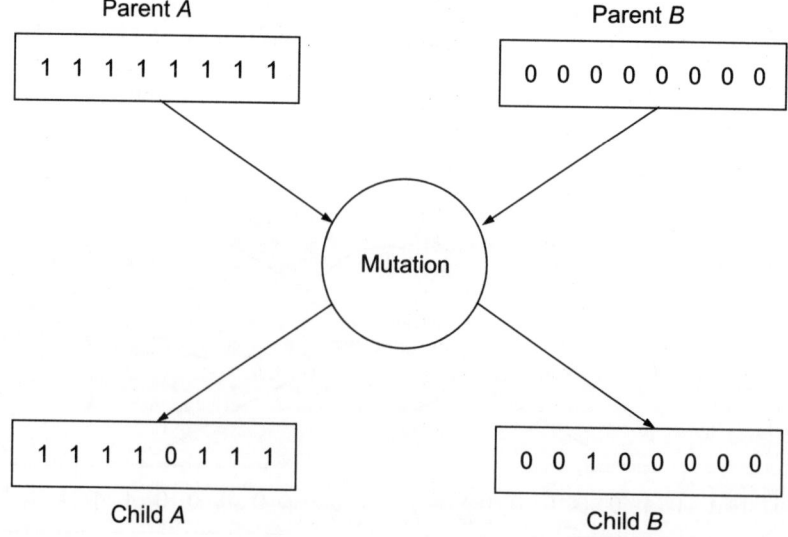

FIGURE 13.9 Typical mutation operation in genetic algorithm.

When the members of the population become similar, there is a loss in the ability to search. Hence, the algorithm is terminated if the average fitness of the population is near to the maximum fitness of any member of the population.

Genetic algorithm has been widely used as a tool for model parameter estimation using measured values of the state variables.

As an example for using GA technique, we present a modelling example, where we model the measurements of HIV-1 viral load for two patients, using a linear fractional order differential equation of the form

$$D_t^a y(t) - ay(t) = 0, \quad y(0) = y_0$$

whose analytical solution is given by

$$y = y_0 E_{\alpha,1}(at^\alpha)$$

where, $E_{\alpha,1}$ is the Mittag–Leffler function defined by

$$E_{\alpha,1}(z) = \sum_{k=0}^{\infty} \frac{z^k}{\Gamma(\alpha k + 1)}$$

and the parameters to be estimated are α, a, and y_0. The parameter estimation is performed using genetic algorithm.

Figure 13.10 shows the actual HIV-1 viral load data points and the computed fit, for the first patient, using the fractional order differential equation, as a function of time in years. Similarly, the measured viral load data and the computed fit for the second patient is shown as a function of time in Figure 13.11. In both the cases, it is seen that the fit is accurate and close to the measurements. Also, high correlation values ($R = 0.9824$ and $R = 0.9923$ for the first and second patient) between the measurements and the computed fit, is observed.

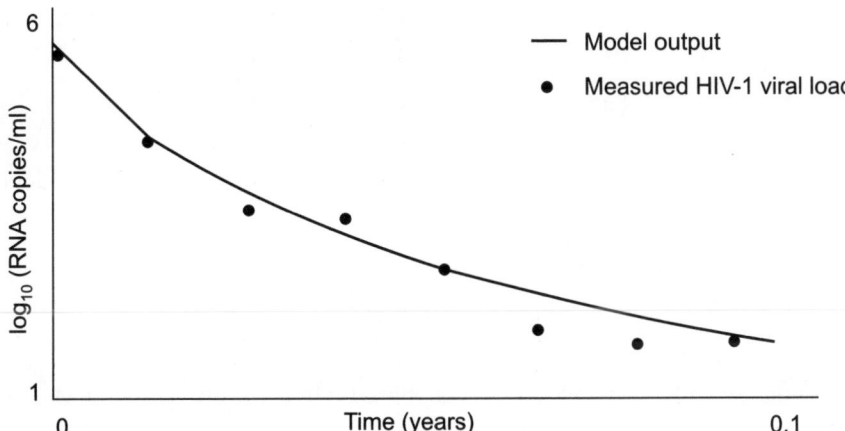

FIGURE 13.10 The measured HIV-1 viral load and the computed fit using Mittag–Leffler solution of the fractional order differential equation shown as a function of time for patient 1.

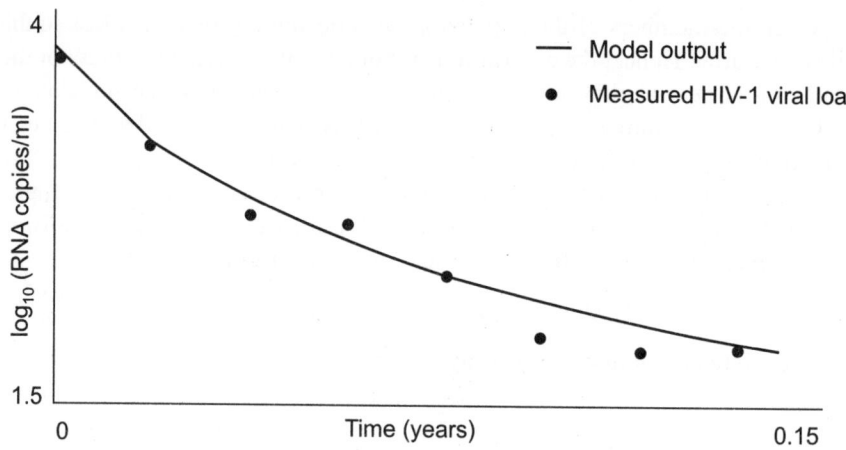

FIGURE 13.11 The measured HIV-1 viral load and the computed fit using Mittag–Leffler solution of the fractional order differential equation shown as a function of time for patient 2.

Further, the applicability of the developed models is illustrated by predicting the viral load at the $(k + 1)$th and $(k + 2)$th instants. It can be seen that the developed models are able to closely predict the values of the viral load at the future instances. Table 13.1 shows the error in prediction of viral load using the developed model. It is observed that the prediction error is low and the developed model, whose parameters were estimated using genetic algorithm, appears to be efficient in representing the dynamics of HIV-1 viral load.

TABLE 13.1 Error in Prediction of the HIV-1 Viral Load at Future Time Instances

	Error in prediction	
	$k + 1$	$k + 2$
Patient 1	0.15	0.04
Patient 2	0.26	0.13

Consider another example where we would like to fit the arterial blood pressure (ABP) measurements given as a function of time to a non-linear fractional order differential equation of the form

$$D^\alpha x(t) = ax + bx^2$$
$$= f(x, t)$$

Using method (see Section 5.11), the solution of the equation can be given as

$$x(t) = h^\alpha f(x, t) - \sum_{m=1}^{\frac{t-\alpha}{h}} (-1)^m \binom{\alpha}{m} x(t - mh)$$

where,

$$\binom{\alpha}{m} = \frac{\Gamma(\alpha+1)}{\Gamma(m+1)\,\Gamma(\alpha-m+1)}$$

The model parameters are estimated using genetic algorithm and the sampling time h is considered to be 0.3 s. The objective function J used is a sum squared error function given as

$$\min_\theta J = \sum_{i=1}^{n} (x_i - \hat{x}_i)^2$$

where, n is the number of data points, $x(t)$ is the measured and tabulated set of arterial blood pressure values with respect to time and $\hat{x}(t)$ is the model output. $\theta = [\alpha \quad a \quad b]$ is the set of model parameter to be estimated. As you can see, in the previous example and also in this case, even the order of the fractional order differential equation is considered as a parameter and can be estimated along with a and b using GA.

Figures 13.12(a) and 13.12(b) show the measured ABP curve and the curve obtained using the model whose parameters are estimated using GA. It is seen that the measured values and the model outputs are close to each other.

Parameter Estimation Algorithms based on Swarm Intelligence

The computational techniques built using the foundations of the intelligent behaviour exhibited in nature such as a school of fish or a swarm of bees are described by a collective term known as *computational intelligence*. The major aim of computational intelligence is to design and develop systems that perform tasks which require intelligence. Such computational techniques based on the intelligent behaviour exhibited in nature are highly useful for solving certain computationally demanding problems.

Swarm intelligence derives its inspiration from the biological examples of swarming and flocking phenomena exhibited by a flock of birds, swarm of bacteria, etc., and consists of several algorithms for solving optimisation problems that cannot be solved using linear search methods. In general, swarm intelligence algorithms mimic the behaviour of the biological creatures within their swarms and colonies. *Swarm intelligence* is the property of a biological system consisting of a group of individuals in which the collective behaviour of the individuals, locally interacting with their environment as well as with each other, causes global intelligent behaviour to emerge. Since swarm intelligence techniques are robust, flexible, fault-tolerant, scalable, and highly parallelisable, they have been widely utilised as methods for model parameter estimation using input-output data (Da San Martino et al., 2006; Kennedy et al., 2001; Consoli et al., 2010). In this section, we present two different swarm intelligence techniques, namely, the particle swarm optimisation (PSO) and the bacterial foraging optimisation.

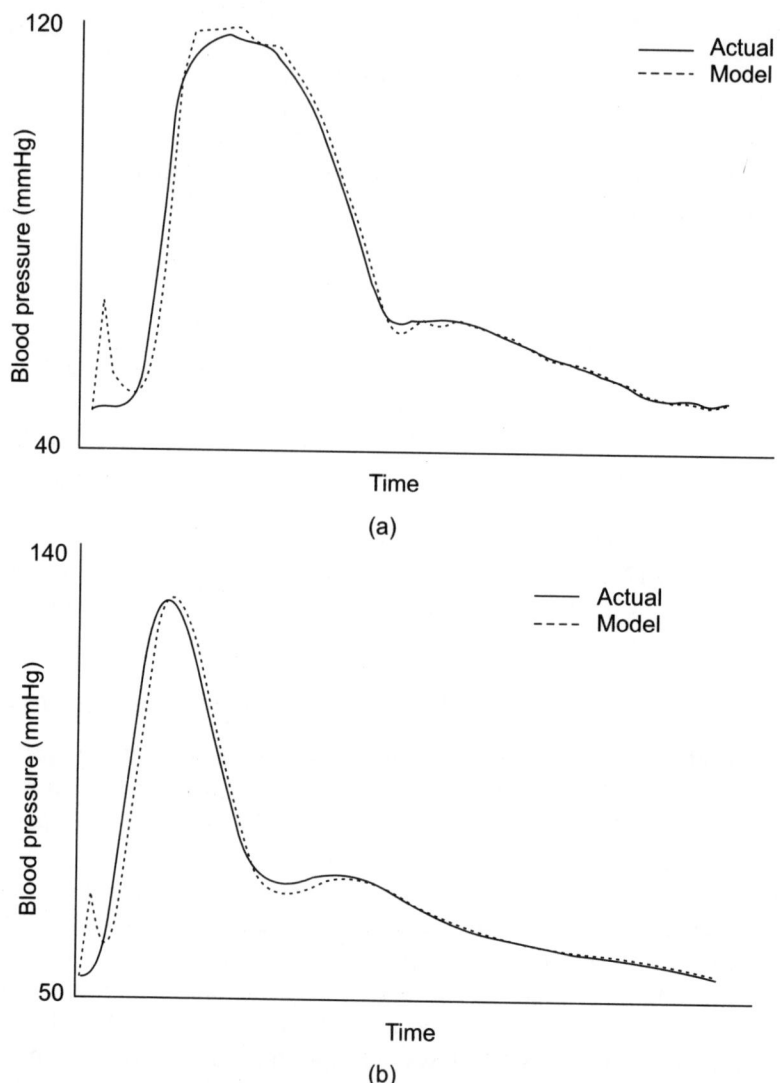

FIGURE 13.12 The measured ABP curve and the curve obtained using the non-linear fractional order differential equation whose parameters are estimated using GA.

Particle swarm optimisation (PSO) technique: The particle swarm optimisation is an efficient swarm intelligence algorithm in which the behaviour of a biological social system like a flock of birds or a school of fish is simulated. For an optimisation problem with d variables, each particle i of the swarm $S = \{1, 2, ..., s\}$ has an associated position vector $x_i = (x_{i1}, x_{i2}, ..., x_{ij}, ..., x_{id})$ and its corresponding velocity vector $v_i = (v_{i1}, ..., v_{ij}, ..., v_{id})$. The coordinates of a particle represent an individual solution of the optimisation problem. The final solution is obtained as the set of trial solutions converging towards the optimal solution due to the foraging group dynamics of the swarm of particles.

Each particle i of the swarm communicates with a subset of S that can vary dynamically. Each particle keeps and uses information of its better position during the process search. Also, it can obtain the best position reached among the particles of its social neighbourhood, which can be the whole swarm or a part of it. The information of the best positions influences the behaviour of particles. In all the cases, the value of the objective function is also stored.

The initial positions and velocities of particles are obtained randomly. In the computer code, the initial set of particles is obtained using a random number generator. At every iteration, the particles update their position and velocity by means of recurrent equations. The position of the particle is modified using exclusively its velocity. However, in updating of its velocity, the value of its own velocity, the best position of the particle (b_i), and the best position of the group of particles of the swarm (g_i) are taken into account. These best positions act with different weights like centres of attraction for particles. The vector equations to update the position x_i and velocity v_i of each particle of the swarm can be described as

$$v_i = c_1 v_i + c_2 \eta (b_i - x_i) + c_3 \eta (g_i - x_i) \tag{13.85}$$

$$x_i = x_{i-1} + v_i \tag{13.86}$$

The parameter c_1 represents the effect of the inertia whose mission is to control the magnitude of the velocity to avoid its indefinite growth. The values c_2 and c_3 are the weights that represent the degree of confidence of the particle. η refers to a random number with uniform distribution in [0, 1] that is independently generated each time (Castro et al., 2009; Kamalanand and Jawahar, 2013).

Algorithm 13.1: Particle swarm optimisation algorithm for estimation of model parameters

Input: Set of available tabulated measurement data of the state variables.
Output: Model parameters.

begin

Step 1: Initialise a population of particles with random positions and velocities on d dimensions in the problem space.

Step 2: For each particle, evaluate the desired optimisation fitness function in d variables.

Step 3: Compare particle's fitness evaluation with particle's pbest. If the current value is better than pbest, then set pbest value equal to current value, and pbest location equal to the current location in d-dimensional space.

Step 4: Compare fitness evaluation with the population's overall previous best. If the current value is better than gbest, then reset gbest to the current particle's array index and value.

Step 5: Change the velocity and position of the particle according to the updated equations:

$$v_i = c_1 v_i + c_2 \eta (b_i - x_i) + c_3 \eta (g_i - x_i)$$

$$x_i = x_{i-1} + v_i$$

Step 6: Loop to step 2 until a stopping criterion (maximum number of iterations of good fitness) is reached.

End

As an example of the PSO technique, the parameter estimation results for the three-dimensional HIV model (see Section 8.8) described by the simultaneous non-linear differential equations given by

$$\frac{dx(t)}{dt} = a(x_0 - x(t)) - bx(t)z(t)$$

$$\frac{dy(t)}{dt} = c(y_0 - y(t)) + dy(t)z(t)$$

$$\frac{dz(t)}{dt} = z(t)(ex(t) - fy(t))$$

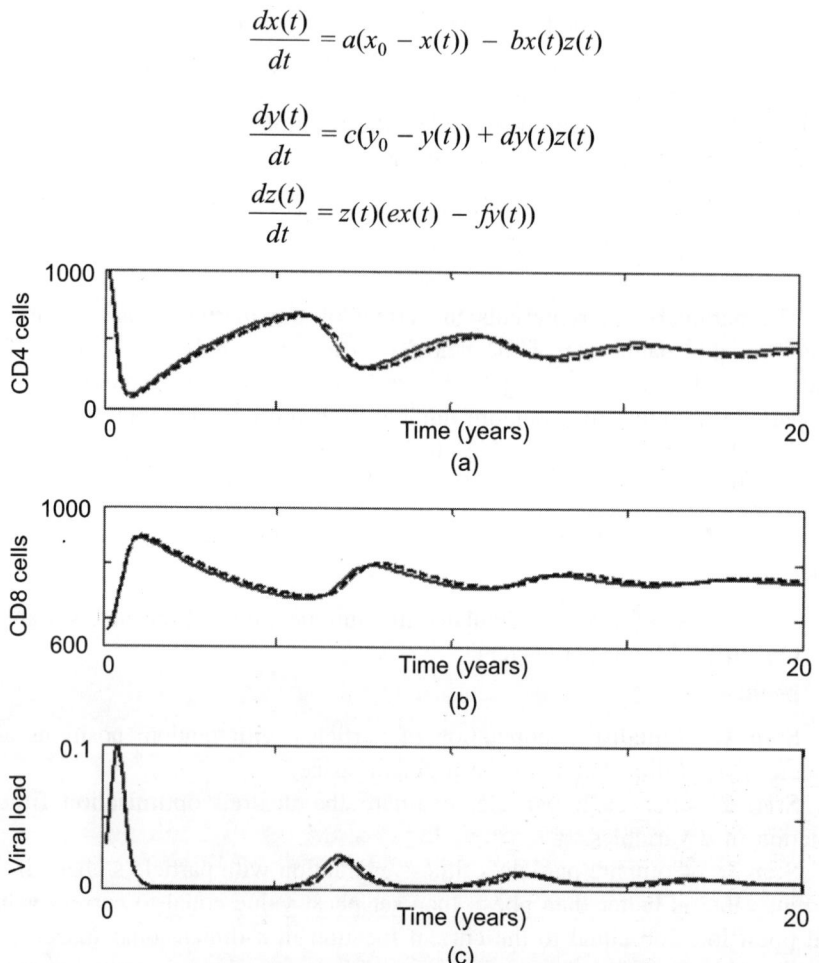

FIGURE 13.13 The measured values (——) and model output (- - - -) values of (a) CD4 cells, (b) CD8 cells and (c) viral load.

For a description of the model, refer to Chapter 8. For estimating all the six parameters of the HIV model using the measurements of CD4 cell count, CD8 cell count and HIV 1 viral load, the value of the parameters c_1, c_2, c_3, c_4, are all set to 0.25. Further, the adopted objective function to be minimised for the estimation of HIV parameters can be described as follows:

$$J_\theta = \sum_{n=1}^{N} \frac{(\hat{x}_n - x_n)^2}{N \text{ mean }(x_n)} + \sum_{k=1}^{K} \frac{(\hat{y}_k - y_k)^2}{K \text{ mean }(y_k)} + \sum_{m=1}^{M} \frac{(\hat{z}_m - z_m)^2}{M \text{ mean }(z_m)}$$

where, $\theta = [a, b, c, d, e, f]$ is the set of HIV parameters to be estimated, x and $\hat{x}$ are the actual and estimated values of CD4 cell population, respectively, y and $\hat{y}$ are the actual and estimated values of CD8 cell population, respectively, z and $\hat{z}$ are the actual and estimated values of viral load respectively, N, K and M are the total number of samples available for CD4, CD8 cell population and viral load, respectively. The estimation is performed using MATLAB 7.0.1 on a computer with 1.73 GHz processor.

Figure 13.13 shows the measured values of CD4, CD8 cell populations and viral load as well as the values obtained using the model whose parameters are estimated using the PSO technique. It is seen that the PSO technique is efficiently able to estimate the parameters of the HIV model when the state variables are available for measurement.

Bacterial foraging optimisation technique: The search and optimal foraging decision making capabilities of the *E. coli* bacteria have been utilised for the development of the swarm intelligence algorithm known as the *bacterial foraging optimisation*. The *E. coli* bacterium has an inherent control mechanism, which enables it to search for food and to avoid harmful or poisonous substances. The selection behaviour of the *E.coli* bacterium is such that it eliminates inferior foraging strategies and improves efficient and successful foraging strategies. This ability of the bacterium is mathematically utilised for obtaining the solution of an optimisation problem.

The foraging dynamics of the bacterium can be described using four stages, namely, swimming and tumbling using flagella, chemotaxis, reproduction and finally, the elimination and dispersal stage. In the first stage, the bacterium moves as a swarm using movements known as swimming and tumbling. Tumbling mode is used to change the future direction of the swim. A chemotactic step is a set of swim steps followed by tumble motion. The environment determines the actual number of swim steps. If the environment has good nutrient concentration in the direction of the swim, then the bacterium uses more swim steps. The end of the chemotactic step is determined by either reaching the maximum number of steps or by reaching a poor environment which has a low nutrient concentration of noxious substances. When the swim steps are stopped, a tumble action takes place.

After chemotaxis, the bacteria population having sufficient nutrients will reproduce and split into two bacteria. Next, an elimination event may occur when the temperature suddenly increases, and therefore, a population of bacteria, which is currently in a region with high concentration of nutrients, is killed. Finally, dispersal of bacteria may occur due to an event such as a sudden flow of water or air, which disperses bacteria from one place to another. The dispersal event can assist chemotaxis, since dispersal may place bacteria near good nutrient sources. However, in some cases, the dispersal event may place the bacteria in a region of harmful substances, thereby causing elimination of a bacterial population.

In mathematical terms, the coordinates of a bacterium represent an individual solution of the optimisation problem. Such a set of trial solutions converges towards the optimal solution (global minima or maxima) due to the foraging group dynamics of the *E. coli* bacteria population.

In order to find the minimum of a fitness function, assume that θ is a coordinate in the solution space and $J(\theta)$ in an attractant-repellant profile, which represents the location of nutrients and noxious substances.

$$J(\theta), \theta \in \mathbb{R}^P \qquad (13.87)$$

$$P(j, k, l) = \{\theta^i(j, k, l) \mid i = 1, 2, ..., S\} \qquad (13.88)$$

Equation (13.88) denotes the positions of each bacterium in the population. For the *i*th bacteria, it represents the bacteria at the *j*th chemotactic step, *k*th reproduction step and the *l*th elimination-dispersal event.

The evaluation of the fitness of each bacteria requires the calculation of attractant and repellant coefficients using Eq. (13.89), and then combined with the fitness value in Eq. (13.90), where J_{cc} represents the combined cell to cell attraction and repelling effects, and $\theta = [\theta_1, \theta_2, ..., \theta_P]^t$. The values d, h and w are the attractant and repellant coefficients.

$$J_{cc}(\theta, P(j, k, l)) = \sum_{m=1}^{S} J_{cc}^i(\theta, \theta^i(j, k, l))$$

$$= \sum_{m=1}^{S} \left[-d_{\text{attractant}} \exp\left(-w_{\text{attractant}} \sum_{m=1}^{p} (\theta_m - \theta_m^i)^2\right) \right]$$

$$+ \sum_{m=1}^{S} \left[-h_{\text{repellant}} \exp\left(-w_{\text{repellant}} \sum_{m=1}^{p} (\theta_m - \theta_m^i)^2\right) \right]$$

$$\qquad (13.89)$$

$$J(\theta) = J(i, j, k, l) + J_{cc}(\theta, P) \qquad (13.90)$$

Each bacterium in the population will tumble once, and the bacteria which have better fitness values after the tumble will go on the swim step under the swim length limit S. The movement can be described by Eq. (13.89), $C(i)$ denotes the step size, and $\phi(j)$ is a random variable in [0, 1]. $C(i) \phi(j)$ is the moving length that the *i*th bacteria take. The evaluation and movement process takes turns to execute until the lifetime of the bacteria reach the limit, and then, the reproduction process takes place. The least healthy bacteria eventually die, while each of the healthier bacteria (those yielding lower value of the objective function) asexually split into two bacteria, which are then placed in the same location.

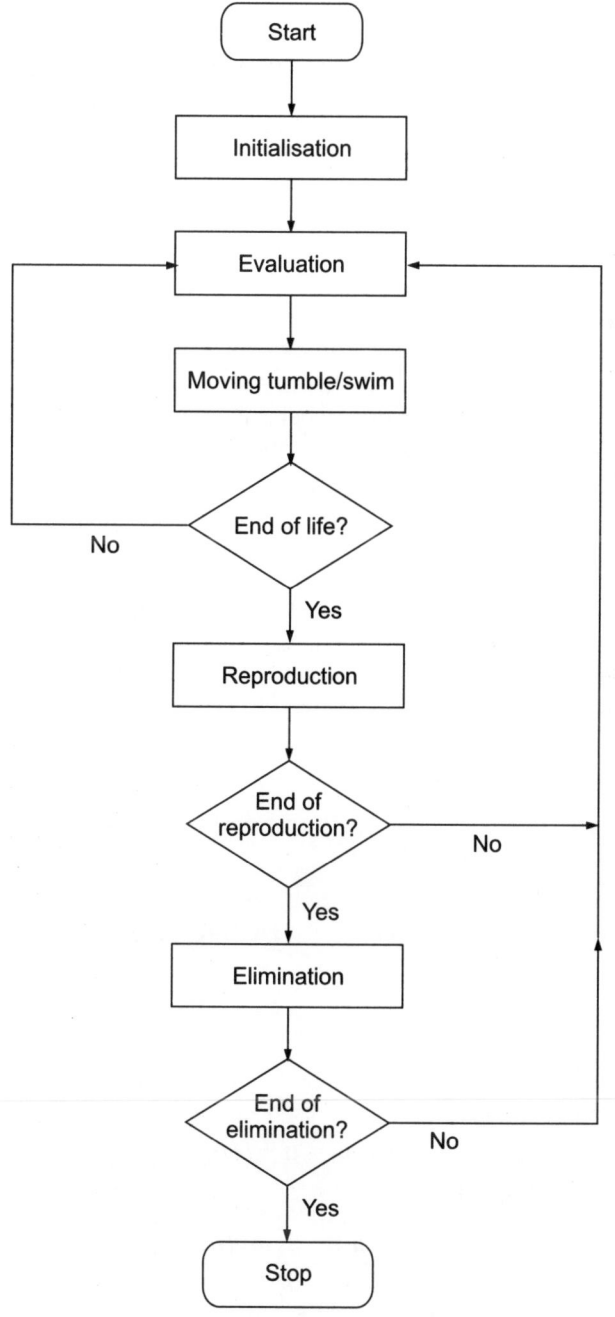

FIGURE 13.14 Flow chart of bacterial foraging optimisation procedure.

$$\theta^i(j+1, k, l) = \theta^i(j, k, l) + C(i)\phi(i) \tag{13.91}$$

Gradual or sudden changes in the local environment may kill a group of bacteria that are currently in a region with high concentration of nutrient gradients. Events can take place in such a fashion that all the bacteria in a region are killed or a group is dispersed into a new location. To simulate this phenomenon in the algorithm, some bacteria are liquidated at random with a very small probability while the new replacements are randomly initialised over the search space (Chen et al., 2007; Korani, 2008). The flow chart of bacterial foraging optimisation procedure is described in Figure 13.14.

13.2.6 Parameter Estimation in Frequency Domain

The estimation of parameters of a system can also be effectively performed in frequency domain. Consider, for example, we would like to model a viscoelastic material as a spring-dashpot system. Few commonly used spring-dashpot representation models of viscoelastic materials are the Voigt model, the Zener model, the Jeffreys model and the fractional Zener model, as shown in Figures 13.15 to 13.18, respectively.

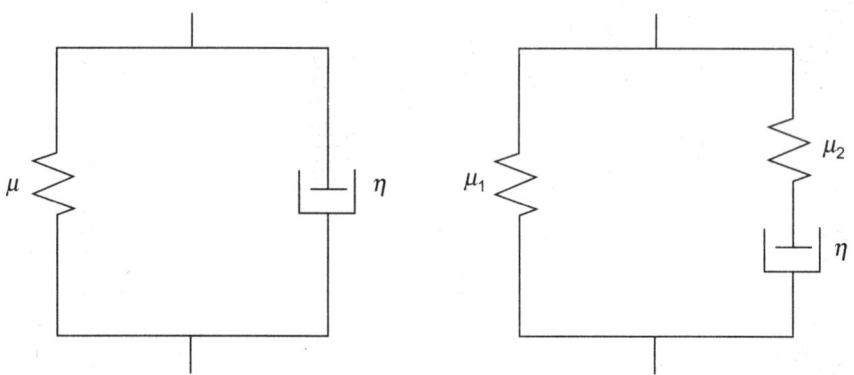

FIGURE 13.15 The Voigt model. **FIGURE 13.16** The Zener model.

where ────◇──── is the springpot. In frequency domain, the model of each system is given as follows:

1. For Voigt model

$$G_M(\omega) = \mu + i\omega\eta$$

2. For Zener model,

$$G_M(\omega) = \frac{\mu_1\mu_2 + i\omega\eta(\mu_1 + \mu_2)}{\mu_2 + i\omega\eta}$$

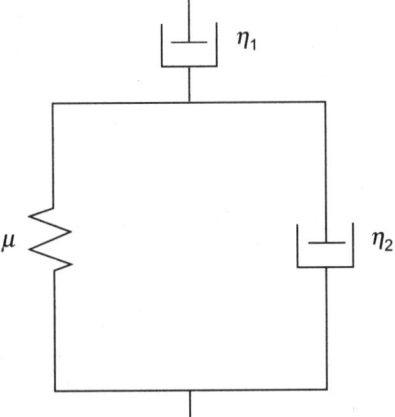

FIGURE 13.17 The Jeffreys model.

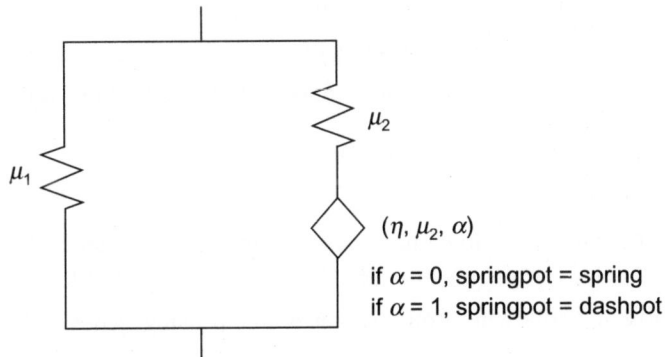

FIGURE 13.18 The fractional Zener model.

3. For Jeffreys model,

$$G_M(\omega) = -\omega\eta_1 \frac{\omega\eta_2 - i\mu}{\mu + i\omega(\eta_1 + \eta_2)}$$

4. For fractional Zener model,

$$G_M(\omega) = \mu_1 + \frac{\mu_2 \left(\frac{i\omega\eta}{\mu_2}\right)^\alpha}{1 + \left(\frac{i\omega\eta}{\mu_2}\right)^\alpha}, \quad 0 \leq \alpha \leq 1$$

$G_M(\omega)$ is the complex modulus and it is equal to $G_1(\omega) + iG_2(\omega)$, where $G_1(\omega)$ is the elastic modulus given as a function of frequency ω and $G_2(\omega)$ is the viscous modulus as a function of frequency ω. Both $G_1(\omega)$ and $G_2(\omega)$ can be measured for various values of ω using a dynamic mechanical analyser (DMA).

After measuring G_1 and G_2, the parameters of the model can be estimated using parameter estimation algorithms by minimising the error E.

$$E = \frac{1}{N}\sum_{n=1}^{N}\sqrt{(\text{Re}[G(\omega_n)] - G_M(\omega_n)])^2 + (\text{Im}[G(\omega_n)] - G_M(\omega_n)])^2}$$

where, $G(\omega_n)$ denotes the measured values, $G_M(\omega_n)$ is the model output and N denotes the number of experimental frequencies.

EXERCISES

1. Derive the parameter update equations of the recursive least squares (RLS) algorithm.
2. Compare unscented Kalman filter with extended Kalman filter and state the advantages of the former over the latter.
3. State Bellman's principle of optimality.
4. Using dynamic programming, maximize the following function:

    max: $F(Y) = y_1^3 + y_2^3 + y_3^3$

    subject to: $y_1 y_2 y_3 \leq 5$, $y_i > 0$ and integers

5. Using a suitable flow chart, describe the use of genetic algorithm for parameter estimation in mathematical modelling problems.
6. Solve the following optimisation problem using genetic algorithm.

    max: $F(x) = x^2$, $0 \leq x \leq 10$

7. Compare particle swarm optimisation with other optimisation algorithms and state its advantages.
8. Write MATLAB programs (particle swarm optimisation and bacterial foraging optimisation) for estimation of all the six parameters of the three-dimensional HIV model using measurements of CD4 cells, CD8 cells and HIV-1 viral load. Obtain the measured values by solving the ODE model for a fixed parameter set and check whether the estimated and true values are close to each other.
9. Develop a hybrid optimisation algorithm using the characteristics of both particle swarm optimisation and bacterial foraging optimisation algorithms.
10. Find out the meaning of patient-specific models and explain their uses in medical diagnostics and treatment planning.

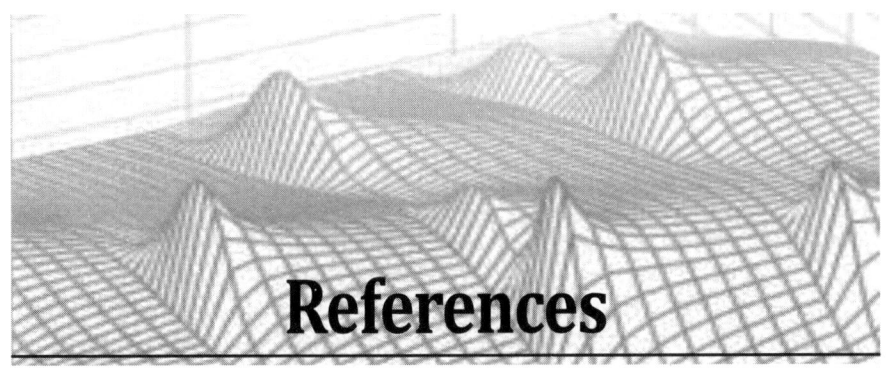

References

Astion, M.L. and Wilding, P. (1992), "Application of Neural Networks to the Interpretation of Laboratory Data in Cancer Diagnosis", *Clinical Chemistry*, **38**(1), 34–38.

Castro, J.P., Landa-Silva, D., and Pérez, J.A.M. (2009), "Exploring Feasible and Infeasible Regions in the Vehicle Routing Problem with Time Windows using a Multi-objective Particle Swarm Optimization Approach", In *Nature Inspired Cooperative Strategies for Optimization (NICSO 2008)*, 103–114, Springer Berlin Heidelberg.

Chamberlain, N.R. (2010), "HIV Infections and AIDS", Lecture.

Chen, T.C., Tsai, P.W., Chu, S.C., and Pan, J.S. (September, 2007), "A Novel Optimization Approach: Bacterial—GA Foraging", *Innovative Computing, Information and Control, ICICIC'07, Second International Conference* (391–394).

Chillingworth, D.R.J. (1976), *Structural Stability of Mathematical Models: The Role of the Catastrophe Method. Mathematical Modelling*, Butterworths, London.

Consoli, S., Moreno-Pérez, J.A., Darby-Dowman, K., and Mladenović, N. (2010), "Discrete Particle Swarm Optimization for the Minimum Labelling Steiner Tree Problem", *Natural Computing*, **9**(1), 29–46.

Douek, D.C., Picker, L.J., and Koup, R.A. (2003), "T Cell Dynamics in HIV-1 Infection", *Annual Review of Immunology*, **21**(1), 265–304.

Einicke, G.A., and White, L.B. (1999), *Robust Extended Kalman Filtering*, IEEE Transactions on Signal Processing, **47**(9), 2596–2599.

Freedman, D.A. (2009), *Statistical Models: Theory and Practice*, Cambridge University Press, New York.

Ge, S.S., Tian, Z., and Lee, T.H. (2005), "Nonlinear Control of a Dynamic Model of HIV-1", *Biomedical Engineering*, **52**(3), 353–361.

Glasstone, S. and Lewis, D. (1983), *Elements of Physical Chemistry*. Macmillan Press, London.

Gunn, S.R. (1998), "Support Vector Machines for Classification and Regression", *ISIS Technical Report*, **14**.

Hamilton, W.R. (1834), "On A General Method in Dynamics; by Which the Study of the Motions of All Free Systems of Attracting or Repelling Points is Reduced to the Search and Differentiation of One Central Relation, or Characteristic Function", *Philosophical Transactions of the Royal Society of London*, **124**, 247–308.

Haykin, S. (2004), *Kalman Filtering and Neural Networks* (Vol. 47), John Wiley & Sons.

Hinton, G.E. (1992), "How Neural Networks Learn from Experience", *Scientific American*, **267**(3), 145–151.

Ho, C.Y.F. and Ling, B.W.K. (2010), "Initiation of HIV Therapy", *International Journal of Bifurcation and Chaos*, **20**(04), 1279–1292.

Julier, S.J. and Uhlmann, J.K. (2004), "Unscented Filtering and Nonlinear Estimation", *Proceedings of the IEEE*, **92**(3), 401–422.

Kaiser, J.F. (1990, April), "On a Simple Algorithm to Calculate the Energy of a Signal", In *Acoustics, Speech, and Signal Processing, 1990, ICASSP-90, 1990 International Conference on* (pp. 381–384), IEEE.

Kaiser, J.F. (1993, April), "Some Useful Properties of Teager's Energy Operators", In *Acoustics, Speech, and Signal Processing, 1993, ICASSP-93, 1993 IEEE International Conference on* (Vol. 3, pp. 149–152), IEEE.

Kamalanand, K. and Jawahar, P.M. (2013), "Particle Swarm Optimization based Estimation of HIV-1 Viral Load in Resource Limited Settings", *Afr. J. Microbiol Res*, **7**, 2297–2304.

Kapur, J.N. (1988), *Mathematical Modelling*, New Age International, New Delhi.

Kennedy, J., Kennedy, J.F., Eberhart, R.C., and Shi, Y. (2001), *Swarm Intelligence*, Morgan Kaufmann, San Francisco.

Kirschner, D. (1996), "Using Mathematics to Understand HIV Immune Dynamics", *AMS Notices*, **43**(2).

Korani, W.M., Dorrah, H.T., and Emara, H.M. (December, 2009), "Bacterial Foraging Oriented by Particle Swarm Optimization Strategy for PID Tuning", In *Computational Intelligence in Robotics and Automation (CIRA)*, 445–450.

Levy, J.A. (1994), "Long-term Survivors of HIV Infection", *Hospital Practice (Office ed.)*, **29**(10), 41–44.

Li, X., Zhou, P. and Aruin, A.S. (2007), "Teager–Kaiser Energy Operation of Surface EMG Improves Muscle Activity Onset Detection", *Annals of Biomedical Engineering*, **35**(9), 1532–1538.

Menezes Campello de Souza, F. (1999), "Modeling the Dynamics of HIV-1 and CD4 and CD8 Lymphocytes", *Engineering in Medicine and Biology Magazine, IEEE*, **18**(1), 21–24.

Neumaier, A. (2003), *Mathematical Modeling*, Institut fur Mathematik, Universitat Wien, http://www.mat.univie.ac.at/neum/ms/model.pdf.

Perelson, A.S., Kirschner, D.E., and De Boer, R. (1993), "Dynamics of HIV Infection of CD4+ T cells", *Mathematical Biosciences*, **114**(1), 81–125.

Richardson, L.F. (1960), *Arms and Insecurity: A Mathematical Study of the Causes and Origins of War*, Boxwood Press, Pittsburg.

Simaan, M. and Cruz, J.B. (1975), "Formulation of Richardson's Model of Arms Race from a Differential Game Viewpoint", *The Review of Economic Studies*, **42**(1), 67–77.

Tay, F.E. and Cao, L. (2001), "Application of Support Vector Machines in Financial Time Series Forecasting", *Omega*, **29**(4), 309–317.

Thomas, G.B. and Finney, R.L. (1992), *Calculus and Analytic Geometry*, Addison-Wesley.

Tu, J.V. (1996), "Advantages and Disadvantages of using Artificial Neural Networks Versus Logistic Regression for Predicting Medical Outcomes", *Journal of Clinical Epidemiology*, **49**(11), 1225–1231.

Vapnik, V., Golowich, S.E., and Smola, A. (1996), "Support Vector Method for Function Approximation, Regression Estimation, and Signal Processing", *Advances in Neural Information Processing Systems,* **9**.

Wan, E.A. and Van Der Merwe, R. (2001), *The Unscented Kalman Filter, Kalman Filtering and Neural Networks*, Chapter 7, Edited by Simon Haykin.

Wang, G., Krueger, G.R., and Buja, L.M. (2004), "A Simple Model to Simulate Cellular Changes in the T Cell System Following HIV-1 Infection", *Anticancer Research*, **24**(3A), 1689–1698.

White, H. (1989), "Some Asymptotic Results for Learning in Single Hidden-layer Feedforward Network Models", *Journal of the American Statistical Association*, **84**(408), 1003–1013.

Further Reading

Andersson, L., Jönsson, U., Johansson, K.H., and Bengtsson, J. (1998), "A Manual for System Identification", Laboratory Exercises in System Identification, KF Sigma i Lund AB, Department of Automatic Control, Lund Institute of Technology, Box, 118.

Andrews, J.G. and McLone, R.R. (Eds.) (1976), *Mathematical Modelling*, Butterworths, London.

Bai, E.W. (2002), "A Blind Approach to the Hammerstein–Wiener Model Identification", *Automatica*, **38**(6), 967–979.

Bai, E.W. and Fu, M. (2002), "A Blind Approach to Hammerstein Model Identification", *IEEE Transactions on Signal Processing*, **50**(7), 1610–1619.

Barker, H.A. and Young, P.C. (1985), *Identification and System Parameter Estimation*, **1**, Pergamon Press, Oxford.

Benaroya, H. and Nagurka, M.L. (2009), *Mechanical Vibration: Analysis, Uncertainties and Control*, CRC Press, London.

Bender, E.A. (1978), *An Introduction to Mathematical Modeling*, Dover Publications, Mineola New York.

Bequette, B.W. (2003), *Process Control: Modeling, Design, and Simulation*, Prentice Hall, New Jersey.

Billings, S.A. and Fakhouri, S.Y. (1977), "Identification of Nonlinear Systems using the Wiener Model", *Electronics Letters*, **13**(17), 502–504.

Billings, S.A. and Fakhouri, S.Y. (1979), "Non-linear System Identification using the Hammerstein Model", *International Journal of Systems Science*, **10**(5), 567–578.

Blum, C. and Li, X. (2008), *Swarm Intelligence in Optimization*, 43–85, Springer Berlin Heidelberg.

Bluman, G. and Kumei, S. (2013), *Symmetries and Differential Equations*, **154**, Springer Science and Business Media, New York.

Brito, D.L. (1972), "A Dynamic Model of an Armaments Race", *International Economic Review*, 359–375.

Buck, R.C. (1978), *Advanced Calculus*, McGraw-Hill.

Cannon, R.H. (2003), *Dynamics of Physical Systems*, Dover Publication, New York.

Cao, L. (2003), "Support Vector Machines Experts for Time Series Forecasting", *Neurocomputing*, **51**, 321–339.

Caponetto, R. (2010), "Fractional Order Systems: Modeling and Control Applications", World Scientific, **72**.

Carpinteri, A. and Mainardi, F. (Eds.) (2014), *Fractals and Fractional Calculus in Continuum Mechanics*, **378**, Springer, London.

Cazelles, B. and Chau, N. (1995), "Adaptive Dynamic Modelling of HIV/AIDS Epidemic using Extended Kalman Filter", *Journal of Biological Systems*, **3**(03), 759–768.

Chen, S., Billings, S.A., and Grant, P.M. (1990), "Non-linear System Identification using Neural Networks", *International Journal of Control*, **51**(6), 1191–1214.

Chocholatý, P. (2013), Integro-differential Equations with Time-varying Delay, "Programs and Algorithms of Numerical Mathematics", *Proceedings of 16th Conference*, Institute of Mathematics, 51–56.

Churchill, R.V. (1981), *Operational Mathematics*, McGraw-Hill, New York.

Corning, P. (2003), *Nature's Magic: Synergy in Evolution and the Fate of Humankind*, Cambridge University Press, UK.

Csele, M. (2011), *Fundamentals of Light Sources and Lasers*, John Wiley & Sons, New Jersey.

Dincer, I. and Cengel, Y.A. (2001), "Energy, Entropy and Exergy Concepts and Their Roles in Thermal Engineering, *Entropy*, **3**(3), 116–149.

Dincer, I. and Rosen, M.A. (2013), *Exergy: Energy, Environment and Sustainable Development*, Elsevier, UK.

Edmonds, D. (2001), *Electricity and Magnetism in Biological Systems*, Oxford University Press, New York.

Efimov, A.V.E., Terpigoreva, V.M., and Zolotarev, J.G. (1985), *Mathematical Analysis (Advanced Topics)*, Mir Publishers, Moscow.

Elsgolts, L. (1977), *Differential Equations and the Calculus of Variations*, Mir Publishers, Moscow.

Engelbrecht, A.P. (2006), *Fundamentals of Computational Swarm Intelligence*, John Wiley & Sons.

Evans, L.C. (2012), "An Introduction to Stochastic Differential Equations", American Mathematical Society, **82**.

Fung, Y.C. (2013), *Biomechanics: Mechanical Properties of Living Tissues*, Springer Science and Business Media, New York.

Gabriel, S., Lau, R.W., and Gabriel, C. (1996), "The Dielectric Properties of Biological Tissues: III. Parametric Models for the Dielectric Spectrum of Tissues", *Physics in Medicine and Biology*, **41**(11), 2271.

Gevers, M. and Ljung, L. (1986), "Optimal Experiment Designs with Respect to the Intended Model Application", *Automatica*, **22**(5), 543–554.

Giordano, F.R., Weir, M.D., and Fox, W.P. (1997), *A First Course in Mathematical Modeling*, Brooks, Cole, London, UK.

Hellsen, R.H.A., Angelis, G.Z., Van de Molengraft, M.J.G., De Jager, A.G., and Kok, J.J. (2000), "Grey-box Modeling of Friction: An Experimental Case-study", *European Journal of Control*, **6**(3), 258–267.

Howard, P. (2005), "Modeling with ODE", math.tamu.edu.

Hunter, I.W. and Korenberg, M.J. (1986), "The Identification of Nonlinear Biological Systems: Wiener and Hammerstein Cascade Models", *Biological Cybernetics*, **55**(2–3), 135–144.

Jawahar, P.M., Parameswaran, A.N., Stalin Mathew, V., and Kumar, M.S. (2002), *A Text Book on Engineering Mechanics*, DD Publications, Chennai.

Johansson, R. (1993), *System Modeling and Identification*, Prentice Hall, Englewood Cliffs, N.J.

Jones, M.T. (2008), *Artificial Intelligence: A System Approach*, Laxmi Publications, New Delhi.

Joseph, D.D. (2013), *Fluid Dynamics of Viscoelastic Liquids,* **84**, Springer-Verlag, Berlin Heidelberg.

Juditsky, A., Hjalmarsson, H., Benveniste, A., Delyon, B., Ljung, L., Sjöberg, J., and Zhang, Q. (1995), "Nonlinear Black-box Models in System Identification: Mathematical Foundations", *Automatica*, **31**(12), 1725–1750.

Kamalanand, K. and Mannar Jawahar, P. (2016), "Comparison of Particle Swarm and Bacterial Foraging Optimization Algorithms for Therapy Planning in HIV/AIDS Patients", *International Journal of Biomathematics*, **9**(02), 1650024.

Kamalanand, K. and Jawahar, P.M. (2013), "Unscented Transformation for Estimating the Lyapunov Exponents of Chaotic Time Series Corrupted by Random Noise", *International Journal of Mathematical Sciences*, **7**(4), 1–6.

Kamalanand, K. and Jawahar, P.M. (2015), "Comparison of Swarm Intelligence Techniques for Estimation of HIV-1 Viral Load", *IETE Technical Review*, **32**(3), 118–195.

Kamalanand, K. and Jawahar, P.M. (2015), "Prediction of Human Immunodeficiency Virus-1 Viral Load from CD4 Cell Count Using Artificial Neural Networks", *Journal of Medical Imaging and Health Informatics*, **5**(3), 641–646.

Kamalanand, K. and Jawahar, P.M. (2014), "Hybrid BFPSO Algorithm based Estimation of Optimal Drug Dosage for Antiretroviral Therapy in HIV-1 Infected Patients", *BMC Infectious Diseases*, **14**(S3), E14.

Kamalanand, K. and Jawahar, P.M. (2013), "Mathematical Modelling of Parental Influence on Human Romantic Relationships", *International Journal of Happiness and Development*, **1**(3), 294–301.

Kamalanand, K. and Ramakrishnan, S. (2015), "Effect of Gadolinium Concentration on Segmentation of Vasculature in Cardiopulmonary Magnetic Resonance Angiograms", *Journal of Medical Imaging and Health Informatics*, **5**(1), 147–151.

Kapur, J.N. (1989), *Maximum-entropy Models in Science and Engineering*, Wiley Eastern, New Delhi.

Karapetyants Kh. M. (1978), *Chemical Thermodynamics*, Mir Publishers, Moscow.

Karmeshu (2003), Entropy Measures, Maximum Entropy Principle and Emerging Applications, **119**, Springer-Verlag, Berlin Heidelberg.

Kazanskaya, A.S. and Skoblo, V.A. (1978), *Calculations of Chemical Equilibria: Examples and Problems*, Mir Publishers, Moscow.

Kienitz, J. and Wetterau, D. (2012), *Financial Modelling: Theory, Implementation and Practice with Matlab Source*, John Wiley & Sons.

Kikoin, A. and Kikoin, I. (1978), *Molecular Physics*, Mir Publishers, Moscow.

Klatt, D., Hamhaber, U., Asbach, P., Braun, J., and Sack, I. (2007), "Noninvasive Assessment of the Rheological Behavior of Human Organs using Multifrequency MR Elastography: A Study of Brain and Liver Viscoelasticity", *Physics in Medicine and Biology*, **52**(24), 7281.

Kleeman, L. (January, 1996), "Understanding and Applying Kalman Filtering", *Proceedings of the Second Workshop on Perceptive Systems*, Curtin University of Technology, Perth Western Australia.

Kreyszig, E. and Norminton, E.J. (1993), *Advanced Engineering Mathematics*, **1002**, Wiley, New York.

Krishnamurthy, K. (2012), "A Method for Estimating the Lyapunov Exponents of Chaotic Time Series Corrupted by Random Noise Using Extended Kalman Filter", *Mathematical Modelling and Scientific Computation* (237–244), Springer-Verlag Berlin Heidelberg.

Krishnamurthy, K. and Ponnuswamy, M.J. (2013). "A Mathematical Exploration into Manipulation and Control of a Bifurcative Dreaming Process", *International Journal of Dream Research*, **6**(2), 110–113.

Kuang, Y. (Ed.) (1993), *Delay Differential Equations: With Applications in Population Dynamics*, **191**, Academic Press.

Lakshmikantham, V. and Rama Mohana Rao, M. (1995), *Theory of Integro-differential Equations*, 1, Gordon and Breach Science Publishers.

Liboff, R.L. (1979), *Introduction to The Theory of Kinetic Equations*, Krieger, Huntington, New York.

Ljung, L. (1987), "System Identification: Theory for the User", *PTR Prentice Hall Information and System Sciences Series*, **198**.

Ljung, L. (1995), *System Identification Toolbox: User's Guide*, Mathworks, Natick, M.A.

Ljung, L. (1998), *System Identification*, 163–173, Birkhäuser Boston, USA.

Ljung, L. (2007), "System Identification Toolbox for Use with MATLAB", *Mathworks*, Natick, MA.

Ll-Duwaish, H. and Karim, M.N. (1997), "A New Method for the Identification of Hammerstein Model, *Automatica*, **33**(10), 1871–1875.

Loney, S.L. (1960), *The Elements of Statics and Dynamics*, University Press, Cambridge.

Lotka, A.J. (1910), "Contribution to the Theory of Periodic Reactions", *The Journal of Physical Chemistry*, **14**(3), 271–274.

Lu, L. and Yao, B. (June, 2010), "Experimental Design for Identification of Nonlinear Systems with Bounded Uncertainties", *American Control Conference (ACC)*, 4504–4509.

MacGuire, M.C. (1965), *Secrecy and the Arms Race*, Harvard University Press, Cambridge, MA.

Maltus, T.R. (2013), "An Essay on the Principle of Population", **1**, Cosimo, New York.

Mantas, P.Q. (1999), "Dielectric Response of Materials: Extension to the Debye Model", *Journal of the European Ceramic Society*, **19**(12), 2079–2086.

Medhi, J. (1984), *Stochastic Processes*, Wiley Eastern, New Delhi.

Monje, C.A., Chen, Y., Vinagre, B. M., Xue, D., and Feliu-Batlle, V. (2010), *Fractional-order Systems and Controls: Fundamentals and Applications*, Springer, London.

Narendra, K.S. and Gallman, P.G. (1966), "An Iterative Method for the Identification of Nonlinear Systems using a Hammerstein Model", *Automatic Control,* **11**(3), 546–550.

Nedjah, N. (2006), *Swarm Intelligent Systems* (Vol. 26), L. de Macedo Mourelle (Ed.), Heidelberg, Springer.

Nedjah, N. and de Macedo Mourelle, L. (2007), *Systems Engineering Using Particle Swarm Optimisation*, Nova Publishers, New York.

Nedjah, N., dos Santos Coelho, L., and de Macedo Mourelle, L. (Eds.) (2008), *Quantum Inspired Intelligent Systems,* **121**, Springer-Verlag Berlin Heidelberg.

Panigrahi, B.K., Das, S., Suganthan, P.N., and Dash, S.S. (Eds.) (2010), *Swarm, Evolutionary, and Memetic Computing: First International Conference on Swarm, Evolutionary, and Memetic Computing, SEMCCO*, Chennai, India, December 16–18, **6466**, Springer-Verlag Berlin Heidelberg.

Petráš, I. (2011), "Fractional Derivatives, Fractional Integrals, and Fractional Differential Equations in Matlab", *Engineering Education and Research Using MATLAB*, InTech, kap, **10**, 239–264.

Petras, I. (2011), *Fractional-order Nonlinear Systems: Modeling, Analysis and Simulation,* Springer-Verlag Berlin Heidelberg.

Phillips, S.J., Anderson, R.P., and Schapire, R.E. (2006), "Maximum Entropy Modeling of Species Geographic Distributions", *Ecological Modelling,* **190**(3), 231–259.

Pintelon, R. and Schoukens, J. (2012), *System Identification: A Frequency Domain Approach*, John Wiley & Sons, New Jersey.

Poole, D.L. and Mackworth, A.K. (2010), *Artificial Intelligence: Foundations of Computational Agents*, Cambridge University Press.

Rahimy, M. (2010), "Applications of Fractional Differential Equations", *Applied Mathematical Sciences,* **4**(50), 2453–2461.

Rao, J.S. and Gupta, K. (1999), *Introductory Course on Theory and Practice of Mechanical Vibrations*, New Age International, India.

Ritger, P.D. and Rose, N.J. (1968), *Differential Equations with Applications*, McGraw-Hill, New York.

Rousseau, C., Saint-Aubin, Y., Antaya, H., Ascah-Coallier, I., and Hamilton, C. (2008), Mathematics and Technology, Springer, New York.

Schiessel, H., Metzler, R., Blumen, A., and Nonnenmacher, T.F. (1995), "Generalized Viscoelastic Models: Their Fractional Equations with Solutions", *Journal of Physics A: Mathematical and General,* **28**(23), 6567.

Schrödinger, E. (1935), "The Present Status of Quantum Mechanics", *Die Naturwissenschaften*, **23**(48), 1–26.

Sierociuk, D. Podlubny, I., and Petras, I. (2013), "Experimental Evidence of Variable-order Behavior of Ladders and Nested Ladders", *Control Systems Technology*, **21**(2), 459–466.

Simmons, G.F. (1972), *Differential Equations: With Applications and Historical Notes*, **452**, McGraw-Hill, New York.

Simon, D. (2006), *Optimal State Estimation: Kalman, H Infinity, and Nonlinear Approaches*, John Wiley & Sons, New Jersey.

Singh, B. (2017), "Computational Tools and Techniques for Biomedical Signal Processing", 1–360, *IGI Global*, doi:10.4018/978-1-5225-0660-7.

Sjöberg, J., Zhang, Q., Ljung, L., Benveniste, A., Delyon, B., Glorennec, P.Y., Hjalmarsson, H., and Juditsky, A. (1995), "Nonlinear Black-box Modeling in System Identification: A Unified Overview", *Automatica*, **31**(12), 1691–1724.

Smith, D. and Moore, L. (2001), "The SIR Model for Spread of Disease", *Journal of Online Mathematics and its Applications*, **1**(3), 1–9.

Smith, H. (2010), "An Introduction to Delay Differential Equations with Sciences Applications to the Life", *Texts in Applied Mathematics*, **57**.

Söderström, T. and Stoica, P. (1988), *System Identification*, Prentice-Hall.

Strogatz, S.H. (1988), "Love Affairs and Differential Equations", *Mathematics Magazine*, **61**(1), 35.

Strogatz, S.H. (1994), *Nonlinear Dynamics and Chaos: With Applications to Physics, Biology, Chemistry, and Engineering*, Perseus Books, US.

Targ, S. (1976), *Theoretical Mechanics—A Short Course*, Mir Publishers, Moscow.

Tuckwell, H.C. (1995), *Elementary Applications of Probability Theory: With an Introduction to Stochastic Differential Equations*, Chapman and Hall, London.

Van Der Heijden, F., Duin, R., De Ridder, D., and Tax, D.M. (2005), *Classification, Parameter Estimation and State Estimation: An Engineering Approach using MATLAB*, John Wiley & Sons, UK.

Velten, K. (2009), "Mathematical Modeling and Simulation: Introduction for Scientists and Engineers", John Wiley & Sons.

Verhulst, P.F. (1845), "Recherches mathématiques sur la loi d'accroissement de la population", *Nouveaux Mémoires de l'Académie Royale des Sciences et Belles-Lettres de Bruxelles*, **18**, 14–54.

Volterra, V. (1928), "Variations and Fluctuations of the Number of Individuals in Animal Species Living Together", *J. Cons. Int. Explor. Mer*, **3**(1), 3–51.

Wachowiak, M.P., Smolíková, R., Zheng, Y., Zurada, J.M., and Elmaghraby, A.S. (2004), "An Approach to Multimodal Biomedical Image Registration Utilizing Particle Swarm Optimization", *Evolutionary Computation*, **8**(3), 289–301.

Wangersky, P.J. and Cunningham, W.J. (1957), "Time Lag in Prey-predator Population Models", *Ecology*, **38**(1), 136–139.

Welch, G. and Bishop, G. (2006), *An Introduction to The Kalman Filter*, University of North Carolina: Chapel Hill, North Carolina, US.

Wigren, T. (1993), "Recursive Prediction Error Identification using the Nonlinear Wiener Model, *Automatica*, **29**(4), 1011–1025.

Yarin, L.P. (2012), *The Pi-Theorem: Applications to Fluid Mechanics and Heat and Mass Transfer*, **1**, Springer-Verlag Berlin Heidelberg.

Young, P.C. (1969), "Applying Parameter Estimation to Dynamic Systems", **I**; *Control engineering*, **16**(10), 119.

Young, P.C. (2011), *Recursive Estimation and Time-series Analysis: An Introduction for the Student and Practitioner*, Springer-Verlag Berlin Heidelberg.

Yunus, A.C. and Cimbala, J.M. (2006), *Fluid Mechanics: Fundamentals and Applications*, International Edition, 185–201, McGraw Hill.

Zeveke, G., Ionkin, P., Netushil, A.V., and Strakhov, S., [Translated by Kuznetsov, B.] (1979), *Analysis and Synthesis of Electric Circuits*, Mir Publishers, Moscow.

Zhao, W.X., Chen, H.F., and Zheng, W.X. (2010), "Recursive Identification for Nonlinear ARX Systems Based on Stochastic Approximation Algorithm", *Automatic Control*, **55**(6), 1287–1299.

Weblinks

http://www.shannonentropy.netmark.pl/

http://www.qsl.net/g4cnn/units/units.htm

Index

Activation function, 190
Angular velocity, 6
Approximation errors, 70
Artificial neural networks, 189
Asymptotically stable, 174
Asymptotic behaviour, 173
Asymptotic stability, 175
Atomic disorder, 18
Autonomous, 30
　system, 35
　of ODEs, 81
Autoregressive (AR) model, 184
Autoregressive exogenous input (ARX) model, 185
Autoregressive moving average (ARMA) model, 184
Autoregressive moving average exogenous input (ARMAX) model, 185

Back propagation training algorithm, 190
Bacterial foraging optimisation, 208, 225, 229
Bellman's principle of optimality, 214
Bernoulli equation, 21
BIBO stability, 175
Biological processes, 189
Bistable system, 158
Black box models, 67, 184
Boltzmann constant, 17, 18, 64, 195
Boundaries, 1
Bounded input bounded output (BIBO) stability, 174

Box-Jenkins (BJ) model, 186
Brachistochrone, 129
Buckingham's π theorem, 64
Butterfly, 161

Calculus of variations, 139
Canonical momentum, 13
Caputo definition, 95
Catastrophe, 159
　models, 160
Causal, 30
　system, 32
Centre, 147, 151
Chemical reaction, 108
Chemotaxis, 229
Closed system, 2, 29, 98
Cole-Cole model, 61
Competition models, 119
Computational intelligence, 225
Consecutive reaction, 114
Conservative force, 7
Conservative systems, 34, 100, 101
Continuous (Boltzmann) entropy, 196
Continuous model, 70
Controllability matrix, 167
Control systems, 5
Coordinates, 1
Correlation coefficient, 50
Cost function, 218
Critical, 142
　points, 144
Crossover, 221

247

Cusp, 161
Cyclic component, 183
Cyclic thermodynamic process, 9

Damping factor, 85
Dead zone, 39
Debye model, 61
Degrees of freedom, 12, 86
Delay differential equation, 94, 123
Density matrix, 199
Describing function, 45
Descriptive statistics, 49
Dichotomy paradox, 48
Dichotomy principle, 48
Difference equation, 91
Differential calculus, 129
Differential-difference equation, 95
Dimensional analysis, 62
Dimensional homogeneity, 20
Dimensionless form, 21
Direct transmission matrix, 163
Discrete model, 70
Discretisation process, 70
Distributed system, 30, 32
Dynamical systems, 67
Dynamic
  analysis, 72
  programming, 214
  systems, 30, 31

E. coli bacteria, 229
Elastic potential energy, 8
Elitist selection, 219
Elliptic umbilic, 161
Empirical models, 53
Energy, 2, 5, 154
Energy balance laws, 66
Energy of electronic excitation, 9
Engineering approximations, 65
Entropy, 16, 17, 18
Epidemiology, 119
Equilibrium, 18
  point, 150
  state, 2
Euler angles, 20
Euler–Lagrange equation, 13, 140

Euler's equation, 132, 133
Euler's formula, 134
Evolutionary optimisation, 218
Exergy, 10
Experimental conditions, 108
Experimental design, 178
Exponential
  distribution, 218
  growth, 83
Exponentially
  decreasing population, 83, 117
  growing population, 116
Extended Kalman filter (EKF), 208

Family of curves, 74, 133
Feedforward neural network, 190
Filtering, 182
Fitness function, 219
Fixed points, 142
Fold, 161
Force-current analogy, 24
Forced, 30
  system, 33
  vibration, 86
Force-voltage analogy, 24
Fourier transform, 61
Fractional differential, 95
Fractional order differential equations, 95
Fractional order systems, 173
Fractional Zener model, 232, 233
Free damped, 85
Friction, 39
Function, 4
Functional, 129

Gamma space, 17
Gaussian noise, 181
Gaussian process, 96
Gaussian random variable, 97
Genetic algorithm, 208, 219, 225
Gibbs entropy, 194
Gravitational energy, 10
Gravitational potential energy, 7
Grey box models, 67
Grunwald–Letnikov definition, 95

Half stable, 143
Hamiltonian, 13, 14, 101
Hamiltonian equations, 14
Hamiltonian formalism, 13
Hamilton's method, 100, 101
Hamilton's principle, 139
Hammerstein model, 187
Hammerstein–Weiner model, 188, 189
Heroin addiction, 124
Heterogeneous system, 2
HIV-1 viral load, 122
HIV infection, 121
HIV model, 123, 228, 229
Homogeneous, 80
    system, 2
Hooke's Law, 63
Human immunodeficiency virus (HIV), 121
Hybrid systems, 30, 36
Hyperbolic umbilic, 161
Hysteresis, 38

Information, 2
Input matrix, 163
Input multiplicity, 40
Integro-differential equation, 70, 94, 123
Intermolecular energy, 9
Internal energy, 5, 9
Internal stability, 173
Intramolecular energy (chemical energy), 9
Intranuclear energy, 9
Irreversible process, 2, 18
Isolated system, 9

Jacobian, 164, 209
    approach, 41
    matrix, 44
Jeffreys model, 232, 233
Joint entropy, 196
Joint probability mass function, 196

Kalman gain, 210
Kepler's second law, 90
Kinetic energy, 5, 8, 9, 13, 15, 140

Kinetic energy, 101, 156
Kullback–Leibler distance, 199

Lagrange multipliers, 194
Lagrangian, 13, 14, 140
    equations, 13
    formalism, 12
Law of conservation
    of energy, 8, 34, 100
    of mass, 98
Law of parsimony, 67
Least squares method, 202
Liapunov function, 155, 156
Liapunov's direct method, 154
Linear, 30, 80
Linear fractional order differential equation, 223
Linearisation, 41
Linear kernel, 193
Linear models, 184
Linear search methods, 225
Line of best fit, 41
Logistic equation with time delay, 123
Logistic model, 117, 200
Lotka–Volterra model, 117
Lotka–Volterra's predator-prey model, 118
Lumped, 30
    system, 31

Macrostate, 17
Malthusian model, 117
Markov process, 96, 183
Mass balance, 66
Mass–energy equivalence, 98
Material balance, 98
Mathematical
    analogy, 23
    modelling, 3
Maximum
    entropy, 18
    entropy principle, 194, 199
    likelihood estimation/estimator, 216
Mean square error, 219
Memory-free systems, 30
Method of least squares, 83

Index

Microstates, 17
Mittag–Leffler function, 95, 223
Modelling, 3
Moment of inertia, 6
Moving average (MA) process, 183
Multi input multi output (MIMO) systems, 173
Multiple regression, 52
Mutation operation, 222
Mutual information, 199

Natural spline, 56
Negative
  definite, 155
  entropy, 19
  semidefinite, 155
  synergy, 19
Negentropy, 19
Neural networks, 190
Newton's law of gravitation, 63
Node, 145, 150
Non-autonomous system, 30, 35
  of ODES, 81
Non-causal system, 30, 33
Non-conservative system, 34
Non-cyclic process, 9
Non-deterministic models, 96
Non-dimensionalisation, 21, 23
Non-homogeneous ODEs, 80
Non-linear ARX model, 186
Non-linear fractional order differential equation, 224
Non-linear ODEs, 80
Non-linear system, 30, 37
Normal distribution, 194, 217
Normalised equation, 23

Objective function, 218, 219
Observability matrix, 168
Occam's Razor, 67
Offline identification, 178
Online parameter estimation, 202
Open system, 2, 29
Operating points, 163
Opposing reaction, 113
Optimisation, 5, 214

Ordinary least squares, 53
Oscillatory dynamics, 84
Output error (OE) model, 185
Output matrix, 163
Output multiplicity, 40
Over-fitting problem, 192

Parameter estimation, 202
Parasitic non-linearity, 38
Particle swarm optimisation, 208, 225, 226
Persistently exciting inputs, 180
Persistently exciting signals, 180
Phase plane, 145
Physical modelling, 3, 49
Pitch, 20
Polynomial regression, 53
Population dynamics, 94, 116
Positive
  semidefinite, 155
  synergy, 19
Positive definite, 155, 157
  function, 155
Potential energy, 5, 7, 8, 9, 15, 101, 105, 140, 156, 157
Potential function, 157
Power, 5
Predator-prey model, 124
Prediction error, 210
Predictive models, 68
Principle of optimal design of experiments, 178
Process, 2
Prototype model, 3
Pseudo random binary signal, 180
Purely random process, 183

Quadrifolium curve, 77
Quantum systems, 25, 199
Quantum zeno effect, 25

Radiant energy, 10
Rate equations, 109
Reachability, 167
Recombination process, 221

Index **251**

Recursive differential equation, 95
Recursive parameter estimation technique, 202
Regression line, 49, 50
Regularised risk function, 192
Relative entropy, 199
Relay, 39
Rényi entropy, 199
Reversible process, 2, 18
Reynolds number, 23
Richardson model, 82
Riemann–Liouville definition, 95
Roll, 20
Romeo–Juliet model, 80
Rotational motion, 6
Roulette wheel selection, 219, 221

Saddle point, 145, 147, 151
Sampling interval, 70
Sampling period, 70
Saturation, 38
Schrödinger's cat, 25
Seasonal component, 183
Second law of thermodynamics, 18
Shannon's entropy, 195
Sigmoid kernel, 193
Simple harmonic motion, 15
Simple physical model, 72
Simulation, 4
Single input single output system, 172
Singularities, 144, 145
Singular points, 142, 144, 150
SIR model, 119
Spiral, 147, 151
Spline functions, 55
Spline interpolation, 55
Squared error, 225
Stability, 154
    analysis, 5, 44
Stable equilibrium, 143
Stable fixed point, 142, 143
Stable node, 145
Stable spiral, 149
State estimator, 209
State space model, 163, 167, 173
State transition matrix, 169
Static system, 30, 67

Stationary process, 96
Steady state, 150
Stochastic process, 96, 183
Structural risk minimisation principle, 192
Sum squared error, 219
Superposition principle, 36, 37
Support vector machines (SVMs), 192
Swallow tail, 161
Swarm intelligence, 218, 225
Synergetic effect, 19
Synergy, 19
Syntropy, 19
System, 1
System identification, 178
System matrix, 44, 163

Tan sigmoid function, 190
Tautochrone, 104
Taylor's formula, 130
Taylor's series, 144
    approximation, 41
Teager–Kaiser energy, 15
Test for controllability, 167
Thermodynamic
    equilibrium, 18
    probability, 17, 18
Three-level lasing process, 125
Time-invariant, 30
    models, 66
    systems, 34, 35, 173, 178
Time-variant
    models, 66, 202
    systems, 30, 35
Training set, 189
Transfer function, 163
    models, 172, 173
Translational motion, 6
Trend, 183
Trifolium curve, 78
Tsallis entropy, 197

Unforced system, 30, 33
Unscented Kalman filter, 208, 210, 213
Unscented transformation, 211
Unstable equilibrium, 150
Unstable fixed point, 142

Unstable node, 145
Unstable spiral, 149

Validation, 4
Variational problems, 129
Viscoelastic material, 232
Voigt model, 232
Voluntary non-linearity, 38
Von Neumann entropy, 199

Weighted entropy, 196
Weiner model, 188
White box models, 66

White noise, 181
   process, 183
Wiener process, 97
Work, 5, 10

Yaw, 20
Young's modulus, 63
Yule process, 183

Zener model, 232
Zeno paradox, 48
Zeno system, 30, 36